FOREWORD

The Structural Eurocodes are a set of structural design standards, developed by CEN over the last 30 years, to cover the design of all types of structures in steel, concrete, timber, masonry and aluminium. In the UK they are published by BSI under the designations BS EN 1990 to BS EN 1999, each in a number of 'Parts'. Each Part will be accompanied by a National Annex that implements the CEN document and adds certain UK-specific provisions.

This publication was originally developed as a teaching resource for university lecturers and students, although it will also be of interest to practising designers. It offers a general overview of design to the Eurocodes and includes a set of design worked examples for structural elements within a notional building. The original SCI publication (P376) includes a version of the set of examples in which the values of partial factors and other parameters where national choice is allowed are the values recommended within the Eurocode parts. In the present publication, all the examples have been re-worked using values given in the UK National Annexes.

The author of the introductory text is Miss M E Brettle of The Steel Construction Institute. Mr A L Smith and Mr D G Brown of The Steel Construction Institute contributed to the worked examples. The re-working to incorporate the UK National Annex values was carried out by Mr D G Brown.

The worked examples were written or checked by:

Dr A J Bell	University of Manchester
Prof. I Burgess	University of Sheffield
Mr M Cullen	Glasgow Caledonian University
Dr B Davison	University of Sheffield
Dr Y Du	SCI (formerly of University of Birmingham)
Dr L Gardner	Imperial College London
Dr A Kamtekar	University of Birmingham
Dr B Kim	University of Plymouth
Dr D Lam	University of Leeds
Dr L-Y Li	University of Birmingham (formerly of Aston University)
Dr J T Mottram	University of Warwick
Mr L P Narboux	Normacadre (formerly of SCI)
Dr P Palmer	University of Brighton
Dr K M Peebles	University of Dundee
Dr J Rizzuto	Faber Maunsell (formerly of University of Coventry)
Dr M Saidani	University of Coventry
Dr K A Seffen	University of Cambridge
Mr N Seward	University of Wales, Newport
Prof. P R S Speare	City University
Mr M Theofanous	Imperial College London (formerly of SCI)
Dr W Tizani	University of Nottingham

The preparation of P376 was funded by Corus Construction Services and Development, and their support is gratefully acknowledged. The preparation of the present publication was funded by The Steel Construction Institute.

SCI PUBLICATION P387

Steel Building Design:
Worked examples for students

*In accordance with Eurocodes
and the UK National Annexes*

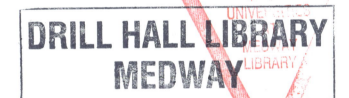

Edited by:

M E Brettle B Eng

Reworked in accordance with the UK National Annexes by
D G Brown C Eng MICE

Published by:
The Steel Construction Institute
Silwood Park
Ascot
Berkshire SL5 7QN

Tel: 01344 636525
Fax: 01344 636570

Publication Number: SCI P387

ISBN 978-1-85942-191-8

British Library Cataloguing-in-Publication Data.

A catalogue record for this book is available from the British Library.

Contents

SUMMARY

This publication offers a general overview of the design of steel framed buildings to the structural Eurocodes and includes a set of worked examples showing the design of structural elements within a notional building. It does not present structural theory or explain detailed design rules. It is intended to be of particular help in undergraduate teaching, although it will also provide guidance to practising designers who want to become acquainted with design to the Eurocodes.

The text discusses the structure of the Eurocode system and the sections contained within a Eurocode Part. It introduces the terminology, and the conventions used for axes and symbols. The document introduces the contents of BS EN 1993 (Eurocode 3) and BS EN 1994 (Eurocode 4) that relate to the design of structural steelwork and steel and composite structures respectively.

The worked examples have all been evaluated using the values of parameters and design options given in the UK National Annexes and are therefore appropriate for structures which are to be constructed in the UK.

The publication has been produced with the assistance of structural design lecturers, who were responsible for writing and checking the majority of the original worked examples presented in Section 6.

1   SCOPE

This publication gives a general overview of structural design to the structural Eurocodes and includes a set of worked examples showing the design of structural elements within a notional building.

The introductory text presents a brief overview of the Eurocodes with respect to the sections, conventions and terminology used. The requirements of BS EN 1993 (steel structures) and BS EN 1994 (composite steel and concrete structures) are briefly introduced with respect to building design. Information is also given for the relevant sections of BS EN 1992 (Eurocode 2), which covers the design of concrete elements in composite structures. Robustness, fire design and corrosion protection are briefly discussed.

The publication has been produced with the assistance of structural design lecturers, who have been responsible for writing and checking the majority of the worked examples presented in Section 6. The set of worked examples present the design of structural elements that may be found in a braced steel framed notional building.

Further design guidance may be found in the documents listed in Section 7 of this publication.

Within the worked examples, frequent reference is made to Access Steel documents. These are a series of publicly available guidance notes on the application of the structural Eurocodes to steelwork design. Many of these notes have the status of non-contradictory complementary information (NCCI), having received endorsement from across Europe. Some notes are UK-specific, relating to UK practice alone. The Access Steel website may be found at www.access-steel.com.

Reference is also made to SCI publication P363, *Steel building design: Design data*. That publication contains comprehensive section property data and member resistances for a wide range of steel sections. The member resistances in P363 have been calculated using the UK National Annexes, and should be directly comparable with the resistances calculated in the present publication. P363 is available from the SCI. Section properties and member resistances are also available from the Corus website (www.corusconstruction.com/en/reference/software).

2 STRUCTURAL EUROCODES SYSTEM

There are ten separate Structural Eurocodes:

EN 1990 Basis of structural design
EN 1991 Actions on structures
EN 1992 Design of concrete structures
EN 1993 Design of steel structures
EN 1994 Design of composite steel and concrete structures
EN 1995 Design of timber structures
EN 1996 Design of masonry structures
EN 1997 Geotechnical design
EN 1998 Design of structures for earthquake resistance
EN 1999 Design of Aluminium Structures

Each Eurocode is comprised of a number of Parts, which are published as separate documents. Each Part consists of:

- Main body of text
- Normative annexes   } These form the full text of the Eurocode Part
- Informative annexes

The full text of each Eurocode Part is issued initially by CEN in three languages with the above 'EN' designation. The full text is then provided with a front cover by each national standards body and published within that country using a designation with the national prefix – for example EN 1990 is published by BSI as BS EN 1990. The Eurocode text may be followed by a National Annex (see Section 2.1 below) or a National Annex may be published separately.

As this set of worked examples are for use in the UK, the full BS EN designation has generally been adopted in the text.

Thus the information contained in the full text of the Eurocodes is the same for each country in Europe. Most parts of the structural Eurocodes present the information using Principles and Application Rules. Principles are denoted by the use of a letter P after the clause number e.g. 1.2(3)P, whereas Application Rules do not contain a letter P e.g. 1.2(3). The former must be followed, to achieve compliance; the latter are rules that will achieve compliance with the Principles but it is permissible to use alternative design rules, provided that they accord with the Principles (see BS EN 1990, 1.4(5)).

The general principle adopted in drafting the Eurocodes was that there would be no duplication of Principles or Application Rules. Thus the design basis given in BS EN 1990 applies irrespective of the construction material or the type of structure. For each construction material, requirements that are independent of structural form are given in 'General' Parts, one for each aspect of design, and form-specific requirements (such as for bridges) are given in other Parts (bridge rules are in Parts 2 of the respective material Eurocodes). Therefore, when designing a structure, many separate Eurocode Parts will be required.

The Structural Eurocodes that may be required for the design of a steel and concrete composite building are:

BS EN 1990   Basis of structural design
BS EN 1991   Actions on structures
BS EN 1992   Design of concrete structures
BS EN 1993   Design of steel structures
BS EN 1994   Design of composite steel and concrete structures
BS EN 1997   Geotechnical design
BS EN 1998   Design of structures for earthquake resistance

In addition to references between structural Eurocode Parts, references to other Standards may be given e.g. product standards.

2.1   National Annexes

Within the full text of a Eurocode, national choice is allowed in the setting of some factors and in the choice of some design methods (i.e. the selection of particular Application Rules); the choices are generally referred to as Nationally Determined Parameters (NDP) and these are published in a National Annex.

The National Annex, where allowed in the Eurocode, will:

- Specify which design method to use.

- Specify what value to use for a factor.

- State whether an informative annex may be used.

In addition, the National Annex may give references to resources that contain non-contradictory complimentary information (NCCI) that will assist the designer. Several National Annexes refer to http://www.steel-ncci.co.uk which has been created to contain NCCI and will be updated with additional resources over time.

The guidance given in a National Annex applies to structures that are to be constructed within that country. National Annexes are likely to differ between countries within Europe.

The National Annexes for the country where the structure is to be constructed should always be consulted in the design of a structure.

Within this publication, the values recommended in the UK National Annexes have been used.

2.2 Geometrical axes convention

The convention for member axes and symbols for section dimensions used in the Eurocodes are shown below.

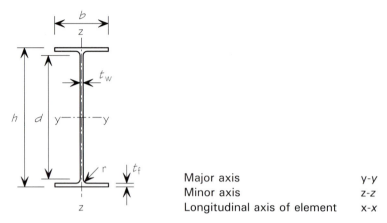

Major axis	y-y
Minor axis	z-z
Longitudinal axis of element	x-x

Figure 2.1 *Axis convention and symbols for principal dimensions*

2.3 Terminology and symbols

The terms used in the Eurocodes have been chosen carefully, for clarity and to facilitate unambiguous translation into other languages. The main terminology used in the Eurocodes includes:

"Actions"   loads, imposed displacements, thermal strain

"Effects"   internal bending moments, axial forces etc.

"Resistance"   capacity of a structural element to resist bending moment, axial force, shear, etc.

"Verification"   check

"Execution"   construction – fabrication, erection

The Structural Eurocodes use the ISO convention for sub-scripts. Where multiple sub-scripts occur, a comma is used to separate them. Four main sub-scripts and their definition are given below:

Eurocode Subscript	Definition	Example	
Ed	Design value of an effect	M_{Ed}	Design bending moment
Rd	Design resistance	M_{Rd}	Design resistance for bending
el	Elastic property	W_{el}	Elastic section modulus
pl	Plastic property	W_{pl}	Plastic section modulus

3 BASIS OF STRUCTURAL DESIGN (BS EN 1990)

BS EN 1990 can be considered as the 'core' document of the structural Eurocode system because it establishes the principles and requirements for the safety, serviceability and durability of structures.

3.1 Limit state design

The information given in the Structural Eurocodes is based on limit state design.

BS EN 1990 defines a limit state as a *'state beyond which the structure no longer fulfils the relevant design criteria'*.

Limit state design provides a consistent reliability against the failure of structures by ensuring that limits are not exceeded when design values of actions, material and product properties, and geotechnical data are considered. Design values are obtained by applying partial factors to characteristic values[1] of actions and properties.

Limit state design considers the resistance, serviceability and durability of a structure. All relevant design situations should be considered for the structure. The design situations considered by the Eurocodes are:

- Persistent – the normal use of the structure.

- Transient – temporary situations, e.g. execution.

- Accidental – exceptional events, e.g. fire, impact or explosion.

- Seismic – seismic events that may act on the structure.

Two limit states are considered during the design process: ultimate and serviceability.

3.1.1 Ultimate limit states

Ultimate limit states are those that relate to the failure of a structural member or a whole structure. Design verifications that relate to the safety of the people in and around the structure are ultimate limit state verifications.

Limit states that should be considered where relevant are:

- Loss of equilibrium of the structure or a structural member.

- Failure of the structure or a structural member caused by: excessive deformation causing a mechanism, rupture, loss of stability, fatigue or other time-dependent effects.

---

[1] The term "characteristic value" applies to actions, material properties and geometrical properties and is defined for each in BS EN 1990. Generally, it means a representative value that has a certain (low) probability of being exceeded (where a greater value would be more onerous) or of not being exceeded (where a lesser value would be more onerous).

- Failure of the supports or foundations, including excessive deformation of the supporting ground.

3.1.2 Serviceability limit states

Serviceability limit states concern the functioning of the structure under normal use, the comfort of the people using the structure and the appearance of the structure. Serviceability limit states may be irreversible or reversible. Irreversible limit states occur where some of the consequences remain after the actions that exceed the limit have been removed, e.g. there is permanent deformation of a beam or cracking of a partition wall. Reversible limit states occur when none of the consequences remain after the actions that exceed the limit have been removed, i.e. the member stresses are within its elastic region.

Criteria that are considered during serviceability limit state design checks are:

- Deflections that affect the appearance of the structure, the comfort of its users and its functionality.

- Vibrations that may cause discomfort to users of the structure and restrict the functionality of the structure.

- Damage that may affect the appearance or durability of the structure.

The Eurocodes do not specify any limits for serviceability criteria, but limits may be given in the National Annexes. The limits should be defined for each project, based on the use of the member and the Client's requirements.

3.2 Combination of actions

BS EN 1990 requires the structure or member to be designed for the critical load cases that are determined by combining actions that can occur simultaneously. This implies that all variable actions that occur concurrently should be considered in a single combination. However, for buildings, note 1 of clause A1.2.1(1) of BS EN 1990 allows the critical combination to be determined from not more than two variable actions. Therefore, engineering judgement may be used to determine the two variable actions that may occur together to produce the critical combination of actions for the whole building or the particular structural member under consideration within the building.

3.2.1 Ultimate limit state

Two methods for determining the combination of actions to be used for the persistent or transient ultimate limit state (ULS) are presented in BS EN 1990. The methods are to use expression (6.10) on its own or, for strength or geotechnical limit states, to determine the least favourable combination from expression (6.10a) and (6.10b). The National Annex for the country in which the building is to be constructed must be consulted for guidance on which method to use – in the UK, either expression (6.10) or the combination of (6.10a) and (6.10b) may be used.

Where multiple independent variable actions occur simultaneously, the Eurocodes consider one to be a leading variable action ($Q_{k,1}$) and the other(s) to be accompanying variable actions ($Q_{k,i}$). A leading variable action is one that has the most onerous effect on the structure or member.

The expressions for the combinations of actions given in BS EN 1990 for ultimate limit state design are shown below.

Persistent or transient design situation

$$\sum_{j\geq1}\gamma_{G,j}G_{k,j} + \gamma_P P + \gamma_{Q,1}Q_{k,1} + \sum_{i>1}\gamma_{Q,i}\psi_{0,i}Q_{k,i} \qquad (6.10)$$

$$\sum_{j\geq1}\gamma_{G,j}G_{k,j} + \gamma_P P + \gamma_{Q,1}\psi_{0,1}Q_{k,1} + \sum_{i>1}\gamma_{Q,i}\psi_{0,i}Q_{k,i} \qquad (6.10a)$$

$$\sum_{j\geq1}\xi_j\gamma_{G,j}G_{k,j} + \gamma_P P + \gamma_{Q,1}Q_{k,1} + \sum_{i>1}\gamma_{Q,i}\psi_{0,i}Q_{k,i} \qquad (6.10b)$$

Accidental design situation

$$\sum_{j\geq1}G_{k,j} + P + A_d + (\psi_{1,1} \text{ or } \psi_{2,1})Q_{k,1} + \sum_{i>1}\psi_{2,i}Q_{k,i} \qquad (6.11b)$$

Seismic design situation

$$\sum_{j\geq1}G_{k,j} + P + A_{Ed} + \sum_{i\geq1}\psi_{2,i}Q_{k,i} \qquad (6.12b)$$

where:

$G_{k,j}$ is the characteristic value of an unfavourable permanent action

P is a prestressing action

$Q_{k,1}$ is the characteristic value of the leading variable action

$Q_{k,i}$ is the characteristic value of an accompanying variable action

A_d is the design value of an accidental action

A_{Ed} is the design value of a seismic action

γ, ψ and ξ are partial, combination and reduction factors on actions, as given in BS EN 1990. These values are subject to modification in the National Annex, which must be consulted.

Typical values of the partial, combination and reduction factors as given in the UK National Annex are given below:

Partial Factor	Permanent action, $\gamma_G = 1.35$
	Variable action, $\gamma_Q = 1.5$
Combination factor	Office areas, $\psi_0 = 0.7$
	Roofs, $\psi_0 = 0.7$
	Snow loads (at lower altitudes), $\psi_0 = 0.5$
	Wind loads, $\psi_0 = 0.5$
Reduction factor	$\xi = 0.925$

Persistent or transient design situation

The combinations of actions given for the persistent or transient design situations are used for static equilibrium, structural resistance and geotechnical design verifications. It should be noted that for structural verification involving geotechnical actions and ground resistance, additional guidance on the approach to determining the combination of actions is given. Annex A of BS EN 1990 presents three different approaches and allows the National Annex to specify which approach to use when considering geotechnical actions. Guidance contained in BS EN 1997 should also be used when considering geotechnical actions.

Accidental design situation

The combination of actions for the accidental design situation can be used to determine a design value that either;

- contains an accidental action (e.g. impact, fire); or

- applies to a situation after an accidental action has occurred (e.g. after a fire).

In the latter case $A_d = 0$.

Seismic design situation

This combination of actions and guidance given in BS EN 1998 should be used when seismic actions are being considered.

3.2.2   Serviceability Limit State

The expressions for the combinations of actions given in BS EN 1990 for serviceability limit state design are shown below.

Characteristic combination	$\sum_{j\geq 1} G_{k,j} + P + Q_{k,1} + \sum_{i>1} \psi_{0,i} Q_{k,i}$	(6.14b)
Frequent combination	$\sum_{j\geq 1} G_{k,j} + P + \psi_{1,1} Q_{k,1} + \sum_{i>1} \psi_{2,i} Q_{k,i}$	(6.15b)
Quasi-permanent combination	$\sum_{j\geq 1} G_{k,j} + P + \sum_{i\geq 1} \psi_{2,i} Q_{k,i}$	(6.16b)

Characteristic combination

This combination of actions should be used when considering an irreversible serviceability limit state. The characteristic combination should be used when considering the functioning of the structure, damage to finishes or non-structural elements.

Frequent combination

Reversible serviceability limit states are covered by the frequent combination of actions. This combination could be used when checking the non-permanent vertical displacement of a floor that supports a machine that is sensitive to vertical alignment.

Quasi-permanent combination

The quasi-permanent combination of actions should be used when considering reversible limit states or long term effects. When considering the appearance of a structure, the quasi-permanent combination should be used.

BS EN 1990 states that advice on which expression (6.14b) to (6.16b) to use is given in the *material* Standard. For steelwork, the National Annex to BS EN 1993 gives suggested limits for calculated vertical deflections and advises that the permanent loads should not be included. The suggested limits are given below.

Vertical deflection	
Cantilevers	length/180
Beams carrying plaster or other brittle finish	Span/360
Other beams (except purlins and sheeting rails)	Span/200
Purlins and sheeting rails	To suit the characteristics of particular cladding

Horizontal deflection limits are also suggested, which are height/300. This limit is not applicable to portal frames.

4   DESIGN PROCESS

The procedures that should be followed when designing a structure are:

1. Choose the structural frame concept, considering:
    *   The layout of the structural members
    *   The type of connections, i.e. simple, semi-rigid or moment resisting
    *   The stability of the structure at all stages (during construction, use and demolition).

2. Determine the actions (loading) on the structure and its members.

3. Analyse the structure, including evaluation of frame stability.

4. Design individual members and connections.

5. Verify robustness.

6. Choose the steel sub-grade.

7. Specify appropriate protection of steel, e.g. against fire and corrosion.

5 BUILDING DESIGN

BS EN 1993-1-1 gives generic design rules for steel structures and specific guidance for structural steelwork used in buildings. It presents design rules for use with the other parts of BS EN 1993 for steel structures and with BS EN 1994 for composite steel and concrete structures.

BS EN 1993-1 comprises twelve parts (BS EN 1993-1-1 to BS EN 1993-1-12). When designing orthodox steel framed buildings, the following parts of BS EN 1993-1 will be required:

BS EN 1993-1-1   General rules and rules for buildings

BS EN 1993-1-2   Structural fire design

BS EN 1993-1-3   Supplementary rules for cold-formed members and sheeting

BS EN 1993-1-5   Plated structural elements

BS EN 1993-1-8   Design of joints

BS EN 1993-1-10   Material toughness and through-thickness properties

When designing a steel and concrete composite building, the following parts of Eurocode 4 will be required:

BS EN 1994-1-1   Design of composite steel and concrete structures - General rules and rules for buildings

BS EN 1994-1-2   Design of composite steel and concrete structures - Structural fire design

In addition to the above, the following Eurocode is needed:

BS EN 1992-1-1   Design of concrete structures - General rules and rules for buildings

5.1 Material properties

5.1.1 Steel grades

The rules in BS EN 1993-1-1 relate to structural steel grades S235 to S460 in accordance with BS EN 10025, BS EN 10210 or BS EN 10219 and thus cover all the structural steels likely to be used in buildings. In exceptional circumstances, components might use higher strength grades; BS EN 1993-1-12 gives guidance on the use of BS EN 1993-1-1 design rules for higher strength steels. For the design of stainless steel components and structures, reference should be made to BS EN 1993-1-4.

Although Table 3.1 of BS EN 1993-1-1 presents steel strengths, the UK National Annex specifies that the nominal yield strength (f_y) and ultimate strength (f_u) of the steel should be taken from the product Standard. The product Standards give more 'steps' in the reduction of strength with increasing thickness of the product. It should be noted that where values from the product standard are used, the specific product standard for the steel grade (e.g. BS EN 10025-2) is required when determining strength values, since there is a

11

slight variation between the Parts of BS EN 10025 for the strength of thicker elements.

The nominal values are used as characteristic values of material strength. Yield and ultimate strength values for common steel thicknesses for S275 and S355 steels given as given in the product Standards are reproduced here in Table 5.1.

The National Annex specifies that when a range of ultimate strengths is given in the product Standard, the lowest value of the range must be chosen. Ultimate strengths in Table 5.1 are therefore the minimum in the quoted range. S275 plate is generally manufactured to BS EN 10025-2. Plate manufactured to other parts of BS EN 10025 may have a slightly different ultimate strength.

Table 5.1   *Yield and ultimate strengths*

Standard and steel grade	Nominal thickness (mm)		
	$t \leq 16$	$16 < t \leq 40$	$3 < t \leq 100$
	Yield strength (f_y) N/mm^2	Yield strength (f_y) N/mm^2	Min. Ultimate strength (f_u) N/mm^2
Sections and plate to BS EN 10025-2			
S275	275	265	410
S355	355	345	470

For yield strengths at thickness > 40mm, consult the Standard
S275 plate is generally manufactured to BS EN 10025-2. Plate manufactured to other parts of BS EN 10025 may have a slightly different ultimate strength.

Although Table 2.1 of BS EN 1993-1-10 can be used to determine the most appropriate steel sub-grade to use, the use of Published Document PD 6695-1-10 for structures to be constructed in the UK is recommended. This PD provides limiting thicknesses related to service temperatures of –5°C and -15°C for internal and exposed steelwork.

5.1.2 Concrete

For structural concrete, BS EN 1994-1-1 refers to BS EN 1992-1-1 for properties but it relates to a narrower range of concrete strength classes than are given in BS EN 1992-1-1 (it omits the lowest and highest grades in BS EN 1992-1-1).

Strength and mechanical properties of concrete for different strength classes are given in Table 3.1 of BS EN 1992-1-1 for normal concrete and in Table 11.3.1 for lightweight aggregate concrete. The concrete strength classes are based on characteristic cylinder strengths (f_{ck}), which are determined at 28 days.

Concrete designations are given typically as C25/30, where the cylinder strength is 25 MPa (N/mm^2) and the cube strength is 30 MPa. Properties are given for a range of lightweight aggregate concrete grades, for densities between 800 and 2000 kg/m^2; a typical designation is LC25/28.

5.1.3 Shear connectors

Properties for headed stud shear connectors should be determined from EN ISO 13918, which covers a range of stud diameters from 10 mm to 25 mm and two materials – structural steel and stainless steel. In determining the design resistance, BS EN 1994-1-1 limits the material ultimate tensile strength to 500 N/mm². When specifying headed stud shear connectors, the designation "SD" is used - for example: "SD 19×100", which is a stud of 19 mm diameter and a nominal height of 100 mm.

5.1.4 Reinforcement

BS EN 1994-1-1, Section 3.2 refers to BS EN 1992-1-1 for the properties of reinforcing steel. However, it should be noted BS EN 1994-1-1 permits the design value of the modulus of elasticity for reinforcing steel to be taken as equal to that for structural steel given in BS EN 1993-1-1 (i.e. 210 kN/mm² rather than 200 kN/mm²).

5.1.5 Profiled steel decking

BS EN 1994-1-1 refers to Sections 3.1 and 3.2 of BS EN 1993-1-3 for the material properties of profiled steel sheeting.

5.2 Section classification

Four classes of cross section are defined in BS EN 1993. Each part of a section that is in compression is classified and the class of the whole cross section is deemed to be the highest (least favourable) class of its compression parts. Table 5.2 of BS EN 1993-1-1 gives limits for the width to thickness ratios for the compression parts of a section for each classification.

The section classification in BS EN 1993-1-1 is adopted for composite sections. Where a steel element is attached to a reinforced concrete element, the classification of the element can, in some cases, be improved. Requirements for ductility of reinforcement in tension are given for class 1 and class 2 cross sections.

5.3 Resistance

Design values of member and connection resistances are determined from characteristic values of material strength and geometrical properties, divided by a partial factor (γ_M). Values of γ_M are given in BS EN 1993-1-1 or BS EN 1994-1-1, as appropriate.

Key values from the National Annexes to BS EN 1993-1-1 and BS EN 1993-1-8 are given below.

Factor	Value
γ_{M0} (resistance of cross-sections)	1.0
γ_{M1} (strength checks of members)	1.0
γ_{M2} (resistance of bolts and welds)	1.25

5.3.1 Cross sectional resistance

Steel sections

Expressions for determining the cross sectional resistance in tension, compression, bending and shear for the four classes of sections are given in Section 6.2 of BS EN 1993-1-1. The design values of resistance are expressed as $N_{t,Rd}$, $N_{c,Rd}$, $V_{c,Rd}$ and $M_{c,Rd}$ respectively.

For slender webs, the shear resistance may be limited by shear buckling; for such situations, reference is made to BS EN 1993-1-5. Shear buckling is rarely a consideration with hot rolled sections.

Composite sections

The design bending resistance of a composite section may be determined by elastic analysis and non-linear theory for any class of cross section; for Class 1 or Class 2 cross-sections, rigid-plastic theory may be used.

Plastic resistance moments of composite sections may be determined either assuming full interaction between the steel and reinforced concrete or for partial shear connection (i.e. when the force transferred to the concrete is limited by the resistance of the shear connectors).

The resistance of a composite section to vertical shear is generally taken simply as the shear resistance of the structural steel section. Where necessary, the resistance of uncased webs to shear buckling should be determined in accordance with BS EN 1993-1-5.

5.3.2 Buckling resistance

Steel sections

Members in compression

BS EN 1993-1-1 presents guidance for checking flexural, torsional and torsional-flexural buckling for members in compression. The Eurocode requires flexural buckling resistance to be verified for all members; torsional and torsional-flexural buckling resistances only need to be verified for members with open cross sections.

A set of five buckling curves is given in Figure 6.4 of BS EN 1993-1-1. The buckling curve is selected appropriate to the cross section type and the axis about which the column buckles. The curves give the value of a reduction factor χ dependent on the non-dimensional slenderness of the member $\bar{\lambda}$. The factor χ is applied as a multiplier to the resistance of the cross section to determine the buckling resistance of the member.

Generally, for columns using hot rolled I and H sections, torsional or torsional-flexural buckling will not determine the buckling resistance of the column.

Members in bending

Laterally unrestrained members in bending about their major axes need to be verified against lateral torsional buckling.

Four buckling curves are defined for lateral torsional buckling, in a similar way to those for flexural buckling of members in compression, but the curves are not illustrated in BS EN 1993-1-1. As for flexural buckling, a reduction factor χ_{LT}

is determined, dependent on the non-dimensional slenderness $\bar{\lambda}_{LT}$ and on the cross section; the rules are given in clause 6.3.2 of BS EN 1993-1-1.

For uniform members in bending, three approaches are given:

- Lateral torsional buckling curves – general case

- Lateral torsional buckling curves for rolled sections and equivalent welded sections

- A simplified assessment method for beams in buildings with discrete lateral restraints to the compression flange.

The second approach gives slightly higher resistances for rolled sections, and is recommended. The UK National Annex should be considered carefully, as it modifies the imperfection factors for rolled sections (affecting tall, narrow beams) and provides specific factors to be used for welded sections.

The guidance given for calculating the beam slenderness for the first two approaches requires the value of the elastic critical moment for lateral torsional buckling (M_{cr}), but no expressions are given for determining this value. Therefore, calculation methods need to be obtained from other sources; three sources are:

- A method for calculating beam slenderness for rolled I, H and channel sections is given in the SCI publication P362 *Steel building design: Concise guide to Eurocode 3*.

- NCCI for calculating M_{cr} is provided on the Access Steel web site (www.access-steel.com).

- *LTbeam*; free software from http://www.cticm.eu/spip.php?lang=en

Members in bending and axial compression

For members subject to bending and axial compression the criteria given in 6.3.3 of BS EN 1993-1-1 must be satisfied.

Interaction factors (k_{ij}) used in the checks may be calculated using either method 1 or 2 given respectively in Annexes A and B of BS EN 1993-1-1. The approach in Annex B is considered to be the simpler of the two methods.

General method for lateral and lateral torsional buckling

The general method given in 6.3.4 of BS EN 1993-1-1 should not be confused with the general case for lateral torsional buckling given in 6.3.2.2 of BS EN 1993-1-1.

The general method gives guidance for structural components that are not covered by the guidance given for compression, bending or bending and axial compression members, and is not likely to be used by most building designers.

Lateral torsional buckling with plastic hinges

Section 6.3.5 of BS EN 1993-1-1 presents guidance for buildings that are designed using plastic analysis, such as portal frames.

5.3.3 Shear Connection

Rules for the verification of the shear connection in composite beams are given in Section 6.6 of BS EN 1994-1-1. Detailed rules are only given for headed stud connectors. Dimension limits and rules for transverse reinforcement are given. Natural bond between the concrete and steel is ignored.

BS EN 1994-1-1 gives the design shear resistance of a headed stud connector as the smaller of the shear resistance of the stud and the crushing strength of the concrete around it. When used with profiled steel sheeting, a reduction factor, based on the geometry of the deck, the height of the stud and the number of studs per trough (for decking perpendicular to the beam), is used to reduce the resistance of the shear connectors.

Limitations are given on the use of partial shear connection, i.e. for situations where the design shear resistance over a length of beam is insufficient to develop the full resistance of the concrete slab.

Longitudinal shear resistance of concrete slabs

The longitudinal shear resistance of a slab is calculated using the procedure given in BS EN 1992-1-1. However, the shear planes that may be critical and the contributions from the reinforcement or the profiled steel sheeting (if the shear connectors are through-deck welded) are defined in BS EN 1994-1-1.

5.4   Joints

BS EN 1993-1-8 gives rules for the design of joints between structural members.
Note that a joint is defined as a zone where two or more members are interconnected; a connection is the location where elements meet and is thus the means to transfer forces and moments.

BS EN 1993-1-8 gives guidance for the design of bolted and welded steel connections subject to predominantly static loading. The steel grades covered are S235, S275, S355 and S460.

BS EN 1993-1-8 classifies joints according to their rotational stiffness as nominally pinned, rigid or semi-rigid. The appropriate type of joint model to be used in global analysis depends on this classification and the method of global analysis. The Standard notes that joints may be classified on the basis of experimental evidence, experience of previous satisfactory performance in similar cases or by calculations based on test evidence. The UK National Annex advises that connections designed in accordance with the principles given in the publication *Joints in steel construction: Simple connections* may be classified as nominally pinned joints.

5.4.1   Bolted connections

BS EN 1993-1-8 defines five categories of bolted connections. These categories distinguish between connections loaded in shear or tension, and connections containing preloaded or non-preloaded bolts. A distinction is also made between preloaded bolts that have slip resistance at serviceability limit state and slip resistance at ultimate limit state. Minimum edge and end distances and bolt spacings are given in terms of the diameter of the bolt hole.

Nominal yield (f_{yb}) and ultimate tensile (f_{ub}) strengths are given for a wide range of bolt classes in Table 3.1 BS EN 1993-1-8; the nominal values should be adopted as characteristic values.

5.4.2 Welded connections

BS EN 1993-1-8 gives guidance for the design of the following types of welds:

- Fillet welds
- Fillet welds all round
- Full penetration butt welds
- Partial penetration butt welds
- Plug welds
- Flare groove welds.

Design resistances of fillet and partial penetration welds are expressed in relation to their throat thickness (rather than leg length) and the ultimate strength of the material joined.

5.5 Robustness

Connections between building members should be designed so that they prevent the building from failing in a manner disproportionate to the event that has caused the structural damage.

BS EN 1991-1-7 gives the design requirements for making structures robust against accidental actions. The Eurocodes separate buildings into 4 classes, with different design requirements for each class of structure.

In addition to the requirements given in the Eurocodes, any national requirements should also be satisfied. In England and Wales, the requirements for the control of disproportionate collapse are given in Approved Document A of the Building Regulations. In Scotland the requirements are given in The Scottish Building Standards, Technical Handbook: Domestic and for Northern Ireland they are given in The Building Regulations (Northern Ireland), Technical Booklet D.

5.6 Fire design / protection

Structural steelwork must either be protected or designed in such a way as to avoid premature failure of the structure when exposed to fire.

Fire protection may be given to structural steelwork members by the use of:

- Intumescent paints
- Mineral boards
- Concrete encasement.

Design guidance for the accidental design situation for fire exposure is given in BS EN 1993-1-2 for structural steelwork and in BS EN 1994-1-2 for composite steel and concrete structures.

5.7   Corrosion protection

The main points to be considered during the design process when deciding on the type of corrosion protection to be applied to the structural steelwork are:

- Application of coating – the need to ensure that the chosen coating can be efficiently applied.

- Contact with other materials.

- Entrapment of moisture and dirt around the steelwork.

- Other factors, e.g. provision of suitable access for maintenance and inspection during the life of the structure.

Types of corrosion protection for structural steelwork members include painted coatings, hot-dip galvanizing and thermal (metal) spraying. Guidance on corrosion protection can be found in the *Corrosion Protection Guides* produced by Corus.

6 WORKED EXAMPLES

The set of worked examples in this Section present the design of structural elements that may be found in a braced steel frame building.

The following should be noted when using the worked examples:

- The structural arrangements used in the notional building considered in this publication are not typical of building design. This is because the structural solutions have been chosen to demonstrate a range of design situations.

- Within the examples, **UK National Annex values have been used.** For construction in other countries, the appropriate National Annexes should be consulted.

- Combination of actions – the examples generally use the least favourable value obtained from either expression (6.10a) or (6.10b) of BS EN 1990, and usually (6.10b), since this is generally the least favourable in orthodox construction.

The worked examples contained in this Section are:

	Job No.		Sheet	*1*	of *2*	Rev	C
	Job Title	Example No. 00		Revised by DGB, April 09			
	Subject	Structural layout and actions					
Silwood Park, Ascot, Berks SL5 7QN Telephone: (01344) 636525 Fax: (01344) 636570	Client		Made by	MEB	Date	Sept 2006	
CALCULATION SHEET			Checked by	DGB	Date	Jan 2008	

Unless stated otherwise all references are to BS EN 1991-1-1:2002	**Structural layout and actions** The various structural arrangements used in the notional building considered in this publication are **not** typical of building design. This is because the structural solutions have been chosen to demonstrate a range of design situations. This example defines the characteristic values of the actions that act on the building shown in Figure 0.1.	

Structural layout and actions

The various structural arrangements used in the notional building considered in this publication are **not** typical of building design. This is because the structural solutions have been chosen to demonstrate a range of design situations.

This example defines the characteristic values of the actions that act on the building shown in Figure 0.1.

Characteristic actions – Floors above ground level

Permanent actions

Self weight of floor	3.5 kN/m^2
Self weight of ceiling, raised floor & services	0.2 kN/m^2
Total permanent action is	

$$g_k = 3.5 + 0.2 = 3.7 \text{ kN/m}^2$$

Permanent action, $g_k = 3.7$ kN/m^2

Variable actions

NA2.4
Table NA.2
Table NA.3
6.3.1.2(8)

Imposed floor load for offices (category B1)	2.5 kN/m^2
Imposed floor load for moveable partitions of less than 2 kN/m run	0.8 kN/m^2
Total variable action is	

$$q_k = 2.5 + 0.8 = 3.3 \text{ kN/m}^2$$

Variable action, $q_k = 3.3$ kN/m^2

Imposed roof actions

Permanent actions

Self weight of roof construction	0.75 kN/m^2
Self weight of ceiling and services	0.15 kN/m^2
Total permanent action is	

$$g_k = 0.75 + 0.15 = 0.9 \text{ kN/m}^2$$

Roof Permanent action, $g_k = 0.9$ kN/m^2

Variable actions

NA 2.10
Table NA.7

The roof is only accessible for normal maintenance and repair

Imposed roof load   0.6 kN/m^2

The imposed roof load due to snow obtained from EN 1991-1-3 is less than 0.6 kN/m^2, therefore the characteristic imposed roof load is taken from EN 1991-1-1.

Roof Variable action, $q_k = 0.6$ kN/m^2

| Example 00 Structural layout and actions | Sheet | 2 | of | 2 | Rev |

BS EN 1991-1-4	## Wind action[0] The total wind force acting on the length of the building (i.e. perpendicular to the ridge) is $F_w = 925$ kN	Wind force acting on the length of the building is: $F_w = 925$ kN

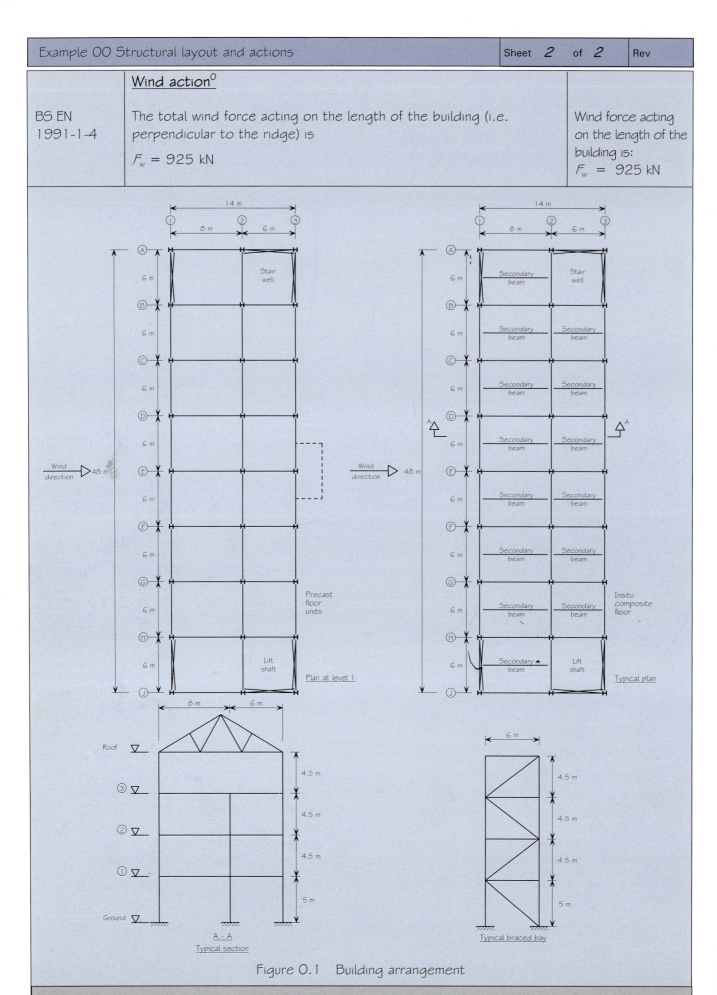

Figure 0.1   Building arrangement

The wind load considered here is only for one direction. Other directions must be considered during the design process. Calculation of the wind loading according to EN 1991-1-4 has not been considered in this example.

	Job No.			Sheet	*1*	of *5*	Rev	C

	Job Title	Example no. 01		Revised by DGB, April 09				
	Subject	Simply supported fully restrained beam						
Silwood Park, Ascot, Berks SL5 7QN Telephone: (01344) 636525 Fax: (01344) 636570	Client		Made by	DL		Date	Nov 2006	
CALCULATION SHEET			Checked by	JTM		Date	Dec 2006	

Unless stated otherwise all references are to BS EN 1993-1-1:2005	## Simply supported fully restrained beam	
	This example demonstrates the design of a fully restrained non-composite beam under uniform loading. The steel beam is horizontal and because the concrete slabs are fully grouted and covered with a structural screed, the compression (top) flange is fully restrained.	
	Consider floor beam at Level 1 – Gridline G1-2	
	Beam span, $L = 8.0$ m Bay width, $= 6.0$ m	
	### Actions	
See structural arrangement and loading	Permanent action Variable action	$g_k = 3.7$ kN/m^2 $q_k = 3.3$ kN/m^2
	### Ultimate limit state (ULS)	
	Partial factors for actions	
BS EN 1990 NA.2.2.3.2 Table NA.A1.2(B)	For the design of structural members not involving geotechnical actions, the partial factors for actions to be used for ultimate limit state design should be obtained from the National Annex.	
	Partial factor for permanent actions $\quad \gamma_G = 1.35$	
	Partial factor for variable actions $\quad \gamma_Q = 1.5$	
	Reduction factor $\quad \xi = 0.925$	
BS EN 1990 6.4.3.2	*Note for this example, the combination factor (ψ_0) is not required as the only variable action is the imposed floor load. The wind has no impact on the design of this member.*	
	Combination of actions at ULS	
BS EN 1990 Eq. (6.10b)	*Design value of combined actions* $= \xi \gamma_G g_k + \gamma_Q q_k$	
	$= (0.925 \times 1.35 \times 3.7) + (1.5 \times 3.3) = 9.57$ kN/m^2	
	UDL per metre length of beam accounting for bay width of 6 m,	ULS design load F_d $= 57.4$ kN/m
	$F_d = 9.57 \times 6.0 = 57.4$ kN/m	
	### Design moment and shear force	
	Maximum design moment, $M_{y,Ed}$, occurs at mid-span, and for bending about the major (y-y) axis is:	Maximum bending moment at mid-span is $M_{y,Ed}$ $= 459$ kNm
	$M_{y,Ed} = \dfrac{F_d L^2}{8} = \dfrac{57.4 \times 8.0^2}{8} = 459$ kNm	

| Example 01 Simply supported fully restrained beam | Sheet | 2 | of | 5 | Rev |

Maximum design shear force, V_{Ed}, occurs at the end supports, and is:

$$V_{Ed} = \frac{F_d L}{2} = \frac{57.4 \times 8}{2} = 230 \text{ kN}$$

Maximum vertical shear force at supports is V_{Ed} = 230 kN

Partial factors for resistance

6.1(1)
NA 2.15

γ_{MO} = 1.0

Trial section

NA 2.4
BS EN 10025-2
Table 7

An Advance UK Beam (UKB) S275 is to be used. Assuming the nominal thickness (t) of the flange and web is less than or equal to 16 mm, the yield strength is:

f_y = 275 N/mm²

Yield strength is f_y = 275 N/mm²

The required section needs to have a plastic modulus about the major-axis (y-y) that is greater than:

$$W_{pl,y} = \frac{M_{y,Ed} \gamma_{MO}}{f_y} = \frac{459 \times 10^3 \times 1.0}{275} = 1669 \text{ cm}^3.$$

From the tables of section properties try section **457 × 191 × 82 UKB, S275**, which has $W_{pl,y}$ = 1830 cm³

P363

Section 457 × 191 × 82 UKB has the following dimensions and properties

Depth of cross-section	h	= 460.0 mm
Web depth	h_w	= 428.0 mm
	(h_w	= $h - 2t_f$)
Width of cross-section	b	= 191.3 mm
Depth between fillets	d	=407.6 mm
Web thickness	t_w	= 9.9 mm
Flange thickness	t_f	= 16.0 mm
Radius of root fillet	r	= 10.2 mm
Cross-sectional area	A	= 104 cm²
Second moment of area (y-y)	I_y	= 37100 cm⁴
Second moment of area (z-z)	I_z	= 1870 cm⁴
Elastic section modulus (y-y)	$W_{el,y}$	= 1610 cm³
Plastic section modulus (y-y)	$W_{pl,y}$	= 1830 cm³

3.2.6(1)

Modulus of elasticity E = 210000 N/mm²

| Example 01 Simply supported fully restrained beam | Sheet *3* | of *5* | Rev |

Classification of cross-section

5.5 & Table 5.2

For section classification the coefficient e is:

$$\varepsilon = \sqrt{\frac{235}{f_y}} = \sqrt{\frac{235}{275}} = 0.92$$

Outstand flange: flange under uniform compression

$$c = \frac{(b - t_w - 2r)}{2} = \frac{(191.3 - 9.9 - 2 \times 10.2)}{2} = 80.5 \text{ mm}$$

$$\frac{c}{t_f} = \frac{80.5}{16.0} = 5.03$$

The limiting value for Class 1 is $\frac{c}{t} \leq 9\varepsilon = 9 \times 0.92 = 8.28$

$5.03 < 8.28$

Therefore, the flange outstand in compression is Class 1.

Internal compression part: web under pure bending

$$c = d = 407.6 \text{ mm}$$

$$\frac{c}{t_w} = \frac{407.6}{9.9} = 41.17$$

The limiting value for Class 1 is $\frac{c}{t} \leq 72\varepsilon = 72 \times 0.92 = 66.24$

$41.17 < 66.24$

Therefore, the web in pure bending is Class 1.

Therefore the section is Class 1 under pure bending.

Section is Class 1

Member resistance verification

6.2.6 *Shear resistance*

6.2.6(1) The basic design requirement is:

$$\frac{V_{Ed}}{V_{c,Rd}} \leq 1.0$$

6.2.6(2)

$$V_{c,Rd} = V_{pl,Rd} = \frac{A_v(f_y / \sqrt{3})}{\gamma_{M0}} \text{ (for Class 1 sections)}$$

6.2.6(3) For a rolled I-section with shear parallel to the web the shear area is

$$A_v = A - 2bt_f + (t_w + 2r)t_f \text{ but not less than } \eta h_w t_w$$

$A_v = 104 \times 10^2 - (2 \times 191.3 \times 16.0) + (9.9 + 2 \times 10.2) \times 16$
$= 4763 \text{ mm}^2$

$\eta = 1.0$ (conservative)

$\eta h_w t_w = 1.0 \times 428.0 \times 9.9 = 4237 \text{ mm}^2$

$4763 \text{ mm}^2 > 4237 \text{ mm}^2$

Therefore, $A_v = 4763 \text{ mm}^2$

25

| Example 01 Simply supported fully restrained beam | Sheet 4 of 5 | Rev |

6.2.6(2)	The design shear resistance is therefore $$V_{c,Rd} = V_{pl,Rd} = \frac{4763 \times (275/\sqrt{3})}{1.0} \times 10^{-3} = 756 \text{ kN}$$ $$\frac{V_{Ed}}{V_{c,Rd}} = \frac{230}{756} = 0.30 < 1.0$$ Therefore, the shear resistance of the section is adequate.	Design shear resistance is: $V_{c,Rd} = 756$ kN Shear resistance is adequate
6.2.6(6)	*Shear buckling* Shear buckling of the unstiffened web need not be considered provided: $$\frac{h_w}{t_w} \leq 72 \frac{\varepsilon}{\eta}$$ $$\frac{h_w}{t_w} = \frac{428.0}{9.9} = 43$$ $$72 \frac{\varepsilon}{\eta} = 72 \times \left(\frac{0.92}{1.0} \right) = 66$$ $43 < 66$ Therefore shear buckling check need not be considered.	
6.2.5(1)	*Moment Resistance* The design requirement is: $$\frac{M_{Ed}}{M_{c,Rd}} \leq 1.0$$	
6.2.5(2)	$$M_{c,Rd} = M_{pl,Rd} = \frac{W_{pl,y} \times f_y}{\gamma_{M0}} \text{ (For Class 1 sections)}$$	
6.2.8(2)	At the point of maximum bending moment the shear force is zero. Therefore the bending resistance does not need to be reduced due to the presence of shear. [1]	
6.2.5(2)	$$M_{c,Rd} = M_{pl,Rd} = \frac{1830 \times 275}{1.0} \times 10^{-3} = 503 \text{ kNm}$$ $$\frac{M_{y,Ed}}{M_{c,Rd}} = \frac{459}{503} = 0.91 < 1.0$$ Therefore, the design bending resistance of the section is adequate.	Design bending resistance is: $M_{c,Rd} = 503$ kNm Bending resistance is adequate

1) Provided that the shear force for the rolled section is less than half of $V_{pl,Rd}$ at the point of maximum bending moment, its effect on the moment resistance may be neglected. At mid-span where the bending moment is at a maximum, the shear force is zero. The maximum shear force occurs at the end supports where for the uniformly distributed load the bending moment is zero. Therefore there is no reduction to the section's design strength, f_y.

| Example 01 Simply supported fully restrained beam | Sheet 5 | of 5 | Rev |

Serviceability Limit State (SLS)

BS EN 1990
NA 2.2.6

Guidance on deflection limits and combinations of actions to be considered are given in the material Standards.

BS EN
1993-1-1
NA 2.23

Vertical deflections should normally be calculated under the characteristic load combination due to variable loads. Permanent loads should not be included.

BS EN1990
6.5.3
(6.14b)

The characteristic load combination at SLS is:

$$\sum G_k + Q_{k,1} + \sum \psi_{0,i} Q_{k,i}$$

This is modified by NA 2.23 to EN 1993-1-1 which states that permanent loads should not be included. As there is only one variable action present, the term $\sum \psi_{0,i} Q_{k,i} = 0$

Vertical deflection of beam

The vertical deflection at the mid-span is determined as:

$$w = \frac{5L^4 q_k}{384 E I_y}$$

$$q_k = 3.3 \times 6.0 = 19.8 \text{ kN/m}$$

$$w = \frac{5 \times 8000^4 \times 19.8}{384 \times 210000 \times 37100 \times 10^4} = 13.6 \text{ mm}$$

Vertical mid-span deflection
$w = 13.6$ mm

BS EN
1993-1-1
NA 2.23

Vertical deflection limit for this example is

$$\frac{\text{Span}}{360} = \frac{8000}{360} = 22.2 \text{ mm}$$

13.6 mm < 22.2 mm

Therefore, the vertical deflection of the section is satisfactory.

Vertical deflection is acceptable

Adopt 457×191×82 UKB in S275 steel

Dynamics

Generally, a check of the dynamic response of a floor beam would be required at SLS. These calculations are not shown here.

Job No.	Sheet *1* of *6*	Rev C
Job Title	Example no. 02	Revised by DGB, April 09
Subject	Simply supported unrestrained beam	

Silwood Park, Ascot, Berks SL5 7QN
Telephone: (01344) 636525
Fax: (01344) 636570

CALCULATION SHEET

Client	Made by	YGD	Date	Nov 2006
	Checked by	DGB	Date	Jan 2008

Unless stated otherwise, all references are to BS EN 1993-1-1

Simply supported unrestrained beam

Introduction

This example demonstrates the design of a simply supported unrestrained beam, as typified by grid line G2-3 on level 1. The beam is 6.0 m long. In this example, it is assumed that the floor slab does not offer lateral restraint. It is also assumed that the loading is not destabilising. In most cases of internal beams if the construction details ensure the load application is not destabilising, it is likely that the details also provide lateral restraint.

Combination of actions at Ultimate Limit State (ULS)

Using the method described in Example 1 the design value of actions for ultimate limit state design is determined as:

$F_d = 60.8$ kN/m

Note: 60.8 kN/m permanent action allows for the self weight of the beam.

Design Values of Bending Moment and Shear Force

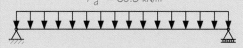

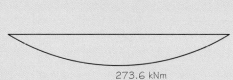

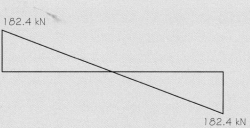

The span of the simply supported beam $L = 6.0$ m

Maximum bending moment at the midspan

$$M_{y,Ed} = \frac{F_d L^2}{8} = \frac{60.8 \times 6^2}{8} = 273.6 \text{ kN/m}$$

Maximum shear force nearby beam support

$$V_{Ed} = \frac{F_d L}{2} = \frac{60.8 \times 6}{2} = 182.4 \text{ kN}$$

Design value of actions
$F_d = 60.8$ kN/m

Design Moment
$M_{Ed} = 273.6$ kNm

Design Shear Force
$V_{Ed} = 182.4$ kN

| Example 02 Simply supported unrestrained beam | Sheet | 2 | of | 6 | Rev |

Partial factors for resistance

6.1(1)
NA 2.15

$\gamma_{M0} = 1.0$

$\gamma_{M1} = 1.0$

Trial section

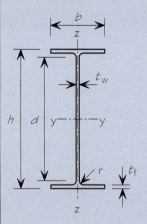

Section Dimensions and Properties of
457 × 191 × 98 UKB, S275

P363

Depth of cross-section	h	= 467.2 mm	
Width of cross-section	b	= 192.8 mm	
Web depth between fillets	d	= 407.6 mm	
Web thickness	t_w	= 11.4 mm	
Flange thickness	t_f	= 19.6 mm	
Root radius	r	= 10.2 mm	
Section area	A	= 125 cm²	
Second moment, y-y	I_y	= 45700 cm⁴	
Second moment, z-z	I_z	= 2350 cm⁴	
Radius of gyration, z-z	i_z	= 4.33 cm	
Warping constant	I_w	= 1180000 cm⁶	
Torsion constant	I_t	= 121 cm⁴	
Elastic section modulus, y-y	$W_{el,y}$	= 1960 cm³	
Plastic section modulus, y-y	$W_{pl,y}$	= 2230 cm³	

Nominal yield strength, f_y of steelwork

NA 2.4
BS EN 10025-2
Table 7

Steel grade = S275,

Flange thickness of the section t_f = 19.6 mm   16 < t_f ≤ 40.0 mm

Hence, nominal yield strength of the steelwork f_y = 265 N/mm²

Yield strength
f_y = 265 N/mm²

Section Classification

Following the procedure outlined in example 1 the cross section under bending is classified as Class 1.

This section is Class 1

Bending Resistance of the cross-section

6.2.5
Eq.6.13

The design resistance of the cross-section for bending about the major axis (y-y) for a class 1 section is:

$$M_{c,Rd} = M_{pl,Rd} = \frac{W_{pl,Rd}f_y}{\gamma_{M0}}$$

| Example 02 Simply supported unrestrained beam | Sheet | 3 | of | 6 | Rev |

$$= \frac{2230 \times 10^3 \times 265}{1.0} \times 10^{-6} = 591 \text{ kNm}$$

Design Bending Resistance
$M_{c,Rd} = 591$ kNm

6.2.5
Eq.6.12

$$\frac{M_{Ed}}{M_{c,Rd}} = \frac{273.6}{591} = 0.46 \ < 1.00 \quad \text{OK}$$

<u>Lateral torsional buckling resistance</u>

Non-dimensional slenderness of an unrestrained beam

6.3.2.2(1)

$$\overline{\lambda}_{LT} = \sqrt{\frac{W_y \times f_y}{M_{cr}}}$$

As BS EN 1993-1-1 does not include an expression for determining M_{cr} an alternative (conservative) method for determining $\overline{\lambda}_{LT}$ is used here.[1]

P 362 Expn (6.55)

$$\overline{\lambda}_{LT} = \frac{1}{\sqrt{C_1}} 0.9 \overline{\lambda}_z \sqrt{\beta_w}$$

P 362 Table 5.5

For a simply supported beam with a uniform distributed load,

$$\frac{1}{\sqrt{C_1}} = 0.94$$

$$\lambda_z = \frac{L}{i_z}$$

$$L = 6000 \text{ mm}^{[2]}$$

$$\lambda_z = \frac{L}{i_z} = \frac{6000}{43.3} = 138.6$$

6.3.1.3

$$\pi \sqrt{\frac{E}{f_y}} = \pi \sqrt{\frac{210000}{265}} = 88.4$$

$$\overline{\lambda}_z = \frac{L}{i_z} \frac{1}{\lambda_1} = \frac{6000}{43.3} \frac{1}{88.4} = 1.568$$

For Class 1 and 2 sections, $\sqrt{\beta_w} = \sqrt{\frac{W_y}{W_{pl,y}}} = \sqrt{\frac{W_{pl,y}}{W_{pl,y}}} = 1.0$

Hence, non-dimensional slenderness

$$\overline{\lambda}_{LT} = \frac{1}{\sqrt{C_1}} 0.9 \overline{\lambda}_z \sqrt{\beta_w} = 0.94 \times 0.90 \times 1.568 \times 1.0 = 1.33$$

slenderness
$\overline{\lambda}_{LT} = 1.33$

1) The calculation of the elastic critical moment (M_{cr}) and thus a less conservative value of $\overline{\lambda}_{LT}$ is given at the end of this example.
2) Conservatively, for a simply supported beam, take the buckling length to equal the span length.

| Example 02 Simply supported unrestrained beam | Sheet | 4 | of | 6 | Rev |

Reduction factor for lateral torsional buckling

6.3.2.3

For rolled I or H section, the reduction factor for torsional buckling

$$\chi_{LT} = \frac{1}{\Phi_{LT} + \sqrt{\Phi_{LT}^2 - \beta\,\overline{\lambda}_{LT}^2}} \quad \text{but} \quad \begin{array}{l} \chi_{LT} \leq 1.00 \\ \chi_{LT} \leq \dfrac{1}{\overline{\lambda}_{LT}^2} \end{array}$$

Where,

$$\Phi_{LT} = 0.5\left[1 + \alpha_{LT}\left(\overline{\lambda}_{LT} - \overline{\lambda}_{LT,0}\right) + \beta\,\overline{\lambda}_{LT}^2\right]$$

6.3.2.3
NA 2.17

The value of $\overline{\lambda}_{LT,0} = 0.4$ (maximum value)

The value of $\beta = 0.75$ (minimum value)

NA 2.17
Table 6.3

For rolled Section with $\dfrac{h}{b} = \dfrac{467.2}{192.8} = 2.42$ and $3.1 \geq 2.42 > 2.0$,

the buckling curve should be **c**, and imperfection factor $\alpha_{LT} = 0.49$

Hence, the value for Φ_{LT} is:

$$\Phi_{LT} = 0.5\,[1 + 0.49 \times (1.33 - 0.4) + 0.75 \times 1.33^2] = 1.391$$

$\Phi_{LT} = 1.391$

Eq.6.57

Reduction factor

$$\chi_{LT} = \frac{1}{1.391 + \sqrt{1.391^2 - 0.75 \times 1.33^2}} = 0.461$$

Check: $\chi_{LT} = 0.461 < 1.00$

$$\chi_{LT} = 0.461 < 1/\overline{\lambda}_{LT}^2 = 1/1.33^2 = 0.565$$

So, reduction factor, $\chi_{LT} = 0.461$

Reduction factor, $\chi_{LT} = 0.461$

Modification of χ_{LT} for moment distribution

NA 2.18
P362

Correction factor due to UDL; $k_c = \dfrac{1}{\sqrt{C_1}} = 0.94$

$$f = 1 - 0.5(1 - k_c)[1 - 2.0(\overline{\lambda}_{LT} - 0.8)^2] \quad \text{but} \leq 1.0$$

$$= 1 - 0.5 \times (1 - 0.94)[1 - 2.0 \times (1.33 - 0.8)^2] = 0.987$$

6.3.2.3
Eq.6.58

Modified reduction factor

$$\chi_{LT,mod} = \frac{\chi_{LT}}{f} = \frac{0.461}{0.987} = 0.467$$

Modified Reduction factor
$\chi_{LT,mod} = 0.467$

Design buckling resistance moment of the unrestrained beam

6.3.2.1
Eq.6.55

$$M_{b,Rd} = \chi_{LT}\frac{W_{pl,y}f_y}{\gamma_{M1}} = 0.467 \times \frac{2230000 \times 265}{1.0} \times 10^{-6} = 276 \text{ kNm}$$

Buckling Resistance
$M_{b,Rd} = 276$ kNm

6.3.2.1
Eq.6.54

$$\frac{M_{Ed}}{M_{b,Rd}} = \frac{274}{276} = 0.99 < 1.0 \quad \text{OK}$$

Buckling resistance adequate

| Example 02 Simply supported unrestrained beam | Sheet | 5 | of | 6 | Rev |

Shear Resistance

6.2.6.3

The shear resistance calculation process is identical to example 1, and is not repeated here.

The calculated shear resistance, $V_{cRd} = 852$ kN, > 182 kN, OK

Adopt 457 × 191 × 98 UKB in S275

Calculation of the elastic critical moment (M_{cr})

Access-steel document SN003a-EN-EU

For doubly symmetric sections, M_{cr} may be determined from:

$$M_{cr} = C_1 \frac{\pi^2 E I_z}{(kL)^2} \left\{ \sqrt{\left[\frac{k}{k_w} \right]^2 \frac{I_w}{I_z} + \frac{(kL)^2 G I_t}{\pi^2 E I_z} + (C_2 z_g)^2} - C_2 z_g \right\}$$

3.2.6(1)

Where:

Modulus of elasticity	E	$= 2100000$ N/mm^2
Shear Modulus	G	$= 81000$ N/mm^2
Distance between lateral supports	L	$= 6000$ mm
No device to prevent beam end from warping	k_w	$= 1$
Compression flange free to rotate about z-z	k	$= 1$

SN003a Table 3.2

For uniformly distributed load on a simply supported beam $\quad C_1 = 1.127$, and $\quad C_2 = 0.454$

z_g is the distance from the load application to the shear centre of the member. When loads applied above the shear centre are destabilising, z_g is positive. Loads applied below the shear centre are not destabilising, and z_g is negative. If loads are not destabilising (as this example), it is conservative to take z_g as zero. When k_w and $k = 1$, and $z_g = $ zero, the expression for M_{cr} becomes:

$$M_{cr} = C_1 \frac{\pi^2 E I_z}{L^2} \left\{ \sqrt{\frac{I_w}{I_z} + \frac{L^2 G I_t}{\pi^2 E I_z}} \right\}$$

$$\frac{\pi^2 E I_z}{L^2} = \frac{\pi^2 \times 210000 \times 23500000}{6000^2 \times 10^3} = 1353 \text{ kN}$$

$$\frac{I_w}{I_z} = \frac{1180000}{2350} = 502.1 \text{ cm}^2$$

$$G I_t = 81000 \times 1210000 \times 10^{-9} = 98.01 \text{ kNm}^2$$

$$M_{cr} = 1.127 \times 1353 \times \left\{ \sqrt{0.05021 + \frac{98.01}{1353}} \right\}$$

$M_{cr} = 534.0$ kNm

$M_{cr} = 534.0$ kNm

6.3.2.2
Eq.6.56

Hence, Non-dimensional slenderness

$$\bar{\lambda}_{LT} = \sqrt{\frac{W_{pl,y} f_y}{M_{cr}}} = \sqrt{\frac{2230000 \times 275}{534.0 \times 10^6}} = 1.07$$

Slenderness

$\bar{\lambda}_{LT} = 1.07$

| Example 02 Simply supported unrestrained beam | Sheet | 6 | of | 6 | Rev |

$\Phi_{LT} = 0.5 [1 + 0.49(1.07 - 0.4) + 0.75 \times 1.07^2] = 1.09$

$$\chi_{LT} = \frac{1}{1.09 + \sqrt{1.09^2 - 0.75 \times 1.07^2}} = 0.601$$

$f = 1 - 0.5 (1 - 0.94)[1 - 2.0(1.07 - 0.8)^2] = 0.974$

$\chi_{LT,mod} = \frac{0.601}{0.974} = 0.617$

$$M_{b,Rd} = \chi_{LT} \frac{W_{pl,y} f_y}{\gamma_{M1}}$$

$$= 0.617 \times \frac{2230000 \times 265}{1.0} \times 10^{-6} = 365 \text{ kNm}$$

This example demonstrates that the simple approach based on

$\bar{\lambda}_{LT} = \frac{1}{\sqrt{C_1}} 0.9 \bar{\lambda}_z \sqrt{\beta_w}$ can produce significant conservatism

compared to the M_{cr} calculation process. (276 kNm compared to 365 kNm)

Serviceability Limit State (SLS) verification

No SLS checks are shown here; they are demonstrated in Example 01.

Buckling Resistance

$M_{b,Rd} = 365$ kNm

<table>
<tr><td rowspan="5">

Silwood Park, Ascot, Berks SL5 7QN
Telephone: (01344) 636525
Fax: (01344) 636570

CALCULATION SHEET</td><td colspan="2">Job No.</td><td colspan="2">Sheet *1* of *10*</td><td>Rev C</td></tr>
<tr><td>Job Title</td><td>Example no. 03</td><td colspan="3">Revised by DGB, April 09</td></tr>
<tr><td>Subject</td><td colspan="4">Simply supported composite secondary beam</td></tr>
<tr><td colspan="2">Client</td><td>Made by</td><td>BK</td><td>Date Nov 07</td></tr>
<tr><td colspan="2"></td><td>Checked by WT</td><td colspan="2">Date Dec 07</td></tr>
</table>

Unless stated otherwise all references are to BS EN 1994-1-1	## Simply supported composite secondary beam

This example shows the design of a 6 m long composite beam subject to UDL, at 3 m centres. The composite slab is 130 mm deep with 1.0 mm gauge *ComFlor 60* (Corus) running perpendicular to the steel beam. The design checks include the moment resistance of the composite beam, the number of shear connectors, vertical shear and transverse reinforcement.

See "Structural arrangement and loading"

Consider the secondary composite beam between ②③ and CD on the typical floor.

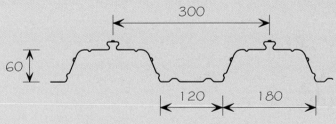

Dimensions of *ComFlor 60* (Corus)

Design data

Beam span	L	= 6.0 m
Beam spacing	s	= 3.0 m
Total slab depth	h	= 130 mm
Depth of concrete above profile	h_c	= 70 mm
Deck profile height	h_p	= 60 mm
Width of the bottom of the trough	b_{bot}	= 120 mm
Width of the top of the trough	b_{top}	= 170 mm approx

Shear connectors

Diameter	d	= 19 mm
Overall height before welding	h_{sc}	= 100 mm
Height after welding		95mm

Materials

BS EN 1993-1-1 NA 2.4
BS EN 10025-2 Table 7

Structural Steel:

For grade S275 and maximum thickness (t) less than 16 mm

Yield strength	f_y	= 275 N/mm²
Ultimate strength	f_u	= 410 N/mm²

BS EN 1992-1-1 Table C.1 BS 4449

Steel reinforcement:

Yield strength	f_{yk}	= 500 N/mm²

| Example 03 Simply supported composite secondary beam | Sheet 2 of 10 | Rev |

BS EN 1992-1-1 Table 3.1

Concrete:

Normal weight concrete strength class C25/30

Density	26 kN/m³ (wet)
	25 kN/m³ (dry)

[These density values may vary for a specific project depending on the amount of steel reinforcement.]

Cylinder strength	f_{ck}	= 25 N/mm²
Secant modulus of elasticity	E_{cm}	= 31 kN/mm²

Actions

Concrete weight

Self weight of the concrete slab (volume from manufacturer's data)

$0.097 \times 26 \times 10^{-6} = 2.52$ kN/m² (wet)

$0.097 \times 25 \times 10^{-6} = 2.43$ kN/m² (dry)

Permanent actions

Construction stage	kN/m²	Composite stage	kN/m²
Steel deck	0.11	Concrete slab	2.43
Steel beam	0.20	Steel deck	0.11
Total	0.31	Steel beam	0.20
		Ceiling and services	0.15
		Total	2.89

Permanent

Construction stage:
$g_k = 0.31$ kN/m²
Composite stage:
$g_k = 2.89$ kN/m²

Variable actions

BS EN 1991-1-6 NA 2.13

Construction stage	kN/m²	Composite stage	kN/m²
Construction loading[1]		Floor load	3.30
(1) Inside and outside the working area	0.75	(See structural arrangement and actions)	
(3) Concrete slab	2.52		
Total	3.27		

Variable

Construction stage:
$q_k = 3.27$ kN/m²
Composite stage:
$q_k = 3.30$ kN/m²

Ultimate Limit State

Combination of actions for Ultimate Limit State

BS EN 1990 Eqn 6.10b NA 2.2.3.2 Table NA A1.2(B)

The design value of combined actions are :

Construction stage:

Distributed load $(0.925 \times 1.35 \times 0.31) + (1.5 \times 3.27) = 5.29$ kN/m²

Total load $F_d = 5.26 \times 6.0 \times 3.0 = 95.2$ kN

Construction stage
$F_d = 95.2$ kN

Composite stage:

Distributed load $(0.925 \times 1.35 \times 2.89) + (1.5 \times 3.3) = 8.56$ kN/m²

Total load $F_d = 8.56 \times 6.0 \times 3.0 = 153.0$ kN

Composite stage
$F_d = 153.0$ kN

---

1) Note that the allowance of 0.75 kN/m² is deemed appropriate in this example, in accordance with NA 2.13 of BS EN 1991-1-6.

| Example 03 Simply supported composite secondary beam | Sheet | 3 | of | 10 | Rev |

Design values of moment and shear force at ULS

Construction stage

Maximum design moment (at mid span)

$$M_{y,Ed} = \frac{F_d L}{8} = \frac{95.2 \times 6.0}{8} = 71.4 \text{ kNm}$$

Composite stage

Maximum design moment (at mid span)

$$M_{y,Ed} = \frac{F_d L}{8} = \frac{153.0 \times 6.0}{8} = 114.8 \text{ kNm}$$

Maximum design shear force (at supports)

$$V_{Ed} = \frac{F_d}{2} = \frac{153.0}{2} = 76.5 \text{ kN}$$

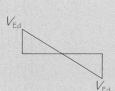

Partial factors for resistance

BS EN
1993-1-1
NA 2.15

| Structural steel | γ_{M0} | = 1.0 |
| Concrete | γ_C | = 1.5 |

BS EN
1992-1-1
NA 2
Table NA.1

NA 2.3
NA 2.4

| Reinforcement | γ_S | = 1.15 |
| Shear connectors | γ_V | = 1.25 |

[Note that the National Annex states that the recommended value of γ_V should be adopted in the absence of more specific information for the type of decking being used. For simplicity, the recommended value of γ_V is used in this example.]

| Longitudinal shear | γ_{Vs} | = 1.25 |

Trial section

The plastic modulus that is required to resist the construction stage maximum design bending moment is determined as:

$$W_{pl,y} = \frac{M_{y,Ed}\gamma_{M0}}{f_y} = \frac{71.4 \times 10^3 \times 1.0}{275} = 259 \text{ cm}^3$$

From the tables of section properties try
section 254 × 102 × 22 UKB, S275, which has $W_{pl,y}$ = 259 cm³

Construction stage:
$M_{y,Ed}$ = 71.4 kNm

Composite stage:
$M_{y,Ed}$ = 114.8 kNm
V_{Ed} = 76.5 kN

| Example 03 Simply supported composite secondary beam | Sheet | 4 | of | 10 | Rev |

P363	Depth of cross-section	h_a	= 254.0 mm
	Width of cross-section	b	= 101.6 mm
	Depth between fillets	d	= 225.2 mm
	Web thickness	t_w	= 5.7 mm
	Flange thickness	t_f	= 6.8 mm
	Radius of root fillet	r	= 7.6 mm
	Cross-section area	A_a	= 28 cm²

[Note the subscript 'a' indicates the steel cross section. A subscript 'c' indicates concrete properties.]

Plastic section modulus (y-y) $W_{pl,y}$ = 259 cm³

BS EN 1993-1-1 NA 2.4

t_f < 16 mm, therefore f_y = 275 N/mm²

BS EN 1993-1-1 3.2.6(1)

Modulus of elasticity E = 210000 N/mm²

Section classification

The section is Class 1 under bending.[2]

Section is Class 1

Note that other construction stage checks are not included in this example.

Composite stage member resistance checks

Concrete

BS EN 1992-1-1 3.1.6 NA 2 Table NA 1

Design value of concrete compressive strength $f_{cd} = \alpha_{cc} \times f_{ck} / \gamma_c$

α_{cc} = 0.85

f_{cd} = 0.85 × 25 / 1.5 = 14.2 N/mm²

f_{cd} = 14.2 N/mm²

Compression resistance of concrete slab

5.4.1.2

At mid-span the effective width of the compression flange of the composite beam is determined from:

$$b_{eff} = b_0 + \sum b_{ei}$$

$$b_{ei} = \frac{L_e}{8} = \frac{L}{8} = \frac{6}{8} = 0.75 \text{ m } (L_e = L \text{ for simply supported beams})$$

Assume a single line of shear studs, therefore, $b_0 = 0$ m

$$b_{eff} = 0 + (2 \times 0.75) = 1.50 \text{ m} < 3 \text{ m (beam spacing)}$$

Effective width b_{eff} = 1.50 m

6.2.1.2

Compression resistance of concrete slab is determined from:

$$N_{c,slab} = f_{cd} b_{eff} h_c$$

where h_c is the depth of the solid concrete above the decking

$$N_{c,slab} = 14.2 \times 1500 \times 70 \times 10^{-3} = 1491 \text{ kN}$$

Design compressive resistance of slab $N_{c,slab}$ = 1491 kN

2) See Example 01 for classification method.

| Example 03 Simply supported composite secondary beam | Sheet 5 | of 10 | Rev |

Tensile resistance of steel section

$$N_{pl,a} = f_d A_a = \frac{f_y A_a}{\gamma_{MO}}$$

$$N_{pl,a} = \frac{275 \times 28 \times 10^2}{1.0} \times 10^{-3} = 770 \text{ kN}$$

Design tensile resistance of steel section
$N_{pl,a}$ = 770 kN

Location of neutral axis

Since $N_{pl,a} < N_{c,slab}$ the plastic neutral axis lies in the concrete flange.

Design bending resistance with full shear connection

6.2.1

As the plastic neutral axis lies in the concrete flange, the plastic resistance moment of the composite beam with full shear connection is:

$$M_{pl,Rd} = N_{pl,a}\left[\frac{h_a}{2} + h - \frac{N_{pl,a}}{N_{c,slab}} \times \frac{h_c}{2}\right]$$

$$M_{pl,Rd} = 770\left[\frac{254}{2} + 130 - \frac{770}{1491} \times \frac{70}{2}\right] \times 10^{-3} = 184 \text{ kNm}$$

Design plastic resistance moment of composite beam
$M_{pl,Rd}$ = 184 kNm

Bending moment at mid span $M_{y,Ed}$ = 114.8 kNm

$$\frac{M_{y,Ed}}{M_{pl,Rd}} = \frac{114.8}{184} = 0.62 < 1.0$$

Therefore, the design bending resistance of the composite beam is adequate, assuming full shear connection.

Design bending resistance is adequate

Shear connector resistance

6.6.3.1

The design shear resistance of a single shear connector in a solid slab is the smaller of:

$$P_{Rd} = \frac{0.29\, \alpha\, d^2\, \sqrt{f_{ck}E_{cm}}}{\gamma_v} \qquad \text{and}$$

$$P_{Rd} = \frac{0.8\, f_u(\pi\, d^2/4)}{\gamma_v}$$

NA 2.3

γ_v = 1.25 for a single stud (see note on sheet 3)

$$\frac{h_{sc}}{d} = \frac{100}{19} = 5.26$$

As $\dfrac{h_{sc}}{d} > 4.0 \qquad \alpha = 1.0$

$$P_{Rd} = \frac{0.29 \times 1.0 \times 19^2\, \sqrt{25 \times 31 \times 10^3}}{1.25} \times 10^{-3} = 73.7 \text{ kN}$$

or

$$P_{Rd} = \frac{0.8 \times 450 \times (\pi \times 19^2/4)}{1.25} \times 10^{-3} = 81.7 \text{ kN}$$

As 73.7 kN < 81.7 kN P_{Rd} = 73.7 kN

| Example 03 Simply supported composite secondary beam | Sheet | 6 | of | 10 | Rev |

Influence of deck shape

6.6.4.2   Deck crosses the beam (i.e. ribs transverse to the beam)

One stud per trough, $n_r = 1.0$

Reduction factor

Eqn 6.23

$$k_t = \left(\frac{0.7}{\sqrt{n_r}}\right)\left(\frac{b_0}{h_p}\right)\left(\frac{h_{sc}}{h_p}-1\right) \le 1.0$$

For trapezoidal profiles, b_0 is the average width of a trough, taken here as $(120 + 170) \div 2 = 145$ mm

$$k_t = \left(\frac{0.7}{\sqrt{1}}\right)\times\left(\frac{145}{60}\right)\times\left(\frac{100}{60}-1\right) = 1.13 \text{ but not more than } 1.0$$

Therefore, as $k_t = 1.0$ no reduction in shear connector resistance is required. Therefore,

$P_{Rd} = 73.7$ kN

Design shear resistance of a single shear stud $P_{Rd} = 73.7$ kN

Number of shear studs in half span

Use one shear connector per trough, therefore,

Stud spacing along beam = 300 mm

Centre line to centre line span of 3 m should be reduced to allow for the primary beam width or the column width (assume 254 mm).

$$n = \frac{3000 - (254/2)}{300} = 9 \text{ stud shear connectors per half span}$$

Provide a stud per trough, total 18 stud shear connectors for the whole span.

Degree of shear connection

Total resistance of 9 shear connectors

$R_q = 9P_{Rd} = 9 \times 73.7 = 663.3$ kN

$$\frac{R_q}{N_{pl,a}} = \frac{663.3}{770} = 0.86 < 1.0$$

$\eta = 0.86$

As this is less than 1.0, this beam has partial shear connection. Therefore, the minimum shear connection requirements must be checked, and the moment resistance reassessed.

| Example 03 Simply supported composite secondary beam | Sheet 7 | of 10 | Rev |

6.6.1.2

Minimum shear connection requirements

The minimum shear connection requirement is calculated from:

(for $L_e < 25m$):

$$\eta \geq 1 - \left(\frac{355}{f_y}\right)(0.75 - 0.03L_e), \quad \eta \geq 0.4$$

For a simply supported beam, L_e is equal to the span.

$$\eta \geq 1 - \left(\frac{355}{275}\right)(0.75 - 0.03 \times 6) = 0.26, \quad \eta \geq 0.4$$

Therefore the degree of shear connection must be at least 0.4. As shown above, there are a sufficient number of shear connectors to achieve this.

6.2.1.3

Design bending moment resistance with partial shear connection

The design bending moment can conservatively be calculated using:

$$M_{Rd} = M_{pl,a,Rd} + \left(M_{pl,Rd} - M_{pl,a,Rd}\right)\eta$$

$M_{pl,a,Rd}$ is the plastic moment resistance of the steel section:

$$M_{pl,a,Rd} = f_{yd}W_{pl,y} = 275 \times 259 \times 10^{-3} = 71.2 \text{ kNm}$$

So the design bending moment resistance is:

$$M_{Rd} = 71.2 + (184 - 71.2) \times 0.86 = 168.2 \text{ kNm}$$

Bending moment at mid span $M_{y,Ed} = 114.8$ kNm

$$\frac{M_{y,Ed}}{M_{Rd}} = \frac{114.8}{168.2} = 0.68 < 1.0$$

Therefore, the design bending resistance of the composite beam with partial shear connection is adequate.

Design bending resistance is adequate

BS EN 1993-1-1 6.2.6(6)

Shear buckling resistance of the uncased web

BS EN 1993-1-5 5.1(2)

For unstiffened webs if $\dfrac{h_w}{t} > \dfrac{72}{\eta}\varepsilon$ the shear buckling resistance of the web should be checked.

Where:

$$\varepsilon = \sqrt{\frac{235}{f_y}} = \sqrt{\frac{235}{275}} = 0.92$$

BS EN 1993-1-1 6.2.6(6)

$\eta = 1.0$ (conservative)

$$h_w = h_a - 2t_f = 254 - (2 \times 6.8) = 240.4 \text{ mm}$$

$h_w = 240.4$ mm

$$\frac{72}{\eta}\varepsilon = \left(\frac{72}{1.0}\right) \times 0.92 = 66.2$$

| Example 03 Simply supported composite secondary beam | Sheet | 8 | of | 10 | Rev |

$$\frac{h_w}{t} = \frac{h_w}{t_w} = \frac{240.4}{5.7} = 42.2$$

As 42.2 < 66.2 the shear buckling resistance of the web does not need to be checked.

Shear buckling resistance check is not required.

Resistance to vertical shear

Vertical shear resistance of the composite beam is:

6.2.2.2

$$V_{pl,Rd} = V_{pl,a,Rd} = \frac{A_v \left(f_y / \sqrt{3} \right)}{\gamma_{M0}}$$

BS EN 1993-1-1 6.2.6(3)

For rolled I and H sections loaded parallel to the web:

$$A_v = A - 2bt_f + t_f \left(t_w + 2r \right) \text{ but not less than } \eta h_w t_w$$

$$A_v = 2800 - (2 \times 101.6 \times 6.8) + 6.8 \times \left[5.7 + (2 \times 7.6) \right]$$

$$A_v = 1560 \text{ mm}^2$$

$\eta = 1.0$ (Conservatively from note to 6.2.6(3))

$$\eta h_w t_w = 1.0 \times 240.4 \times 5.7 = 1370 \text{ mm}^2$$

$$1560 \text{ mm}^2 > 1370 \text{ mm}^2$$

Therefore, $A_v = 1560 \text{ mm}^2$

$$V_{pl,Rd} = 1560 \times \frac{275}{\sqrt{3} \times 1.0} \times 10^{-3} = 247 \text{ kN}$$

Design vertical shear resistance $V_{pl,Rd}$ = 247 kN

$$\frac{V_{Ed}}{V_{pl,Rd}} = \frac{76.5}{247} = 0.31 < 1.0$$

Therefore the design resistance to vertical shear is adequate.

Design resistance for vertical shear is adequate

6.2.2.4

As there is no shear force at the point of maximum bending moment (mid span) no reduction (due to shear) in bending resistance is required.

Design of the transverse reinforcement

For simplicity, neglect the contribution of the decking and check the resistance of the reinforced concrete flange to splitting.

BS EN 1992-1-1 6.2.4 (4)

The area of reinforcement (A_{sf}) can be determined using the following equation:

$$\frac{A_{sf} f_{yd}}{s_f} > \frac{v_{Ed} h_f}{\cot \theta_f} \quad \text{therefore,} \quad \frac{A_{sf}}{s_f} > \frac{v_{Ed} h_f}{f_{yd} \cot \theta_f}$$

Example 03 Simply supported composite secondary beam | Sheet 9 of 10 | Rev

where:

h_f is the depth of concrete above the metal decking, therefore,

6.6.6.4 (1)

$h_f = h_c = 70$ mm

S_f is the spacing of the transverse reinforcement

$$f_{yd} = \frac{f_y}{\gamma_s} = \frac{500}{1.15} = 435 \text{ N/mm}^2$$

For compression flanges $26.5° \le \theta_f \le 45°$

6.6.6.1

The longitudinal shear stress is the stress transferred from the steel beam to the concrete. This is determined from the minimum resistance of the steel, concrete and shear connectors. In this example, with partial shear connection, that maximum force that can be transferred is limited by the resistance of the shear connectors over half of the span, and is given by $R_q = 663.3$ kN

Figure 6.16

This force must be transferred over each half-span. As there are two shear planes (one on either side of the beam, running parallel to it), the longitudinal shear stress is:

$$v_{Ed} = \frac{R_q}{2h_f \Delta x} = \frac{663 \times 1000}{2 \times 70 \times 3000} = 1.58 \text{ N/mm}^2$$

For minimum area of transverse reinforcement assume $\theta = 26.5°$

BS EN 1992-1-1 6.2.4 (4)

$$\frac{A_{sf}}{s_f} \ge \frac{v_{Ed} h_f}{f_{yd} \cot \theta_f} = \frac{1.58 \times 70}{435 \times \cot 26.5} \times 10^3 = 126.7 \text{ mm}^2/\text{m}$$

Therefore, provide A193 mesh reinforcement (193mm²/m) in the slab.[3]

Use A193 mesh reinforcement

Crushing of the concrete compression strut

BS EN 1992-1-1

The model for resisting the longitudinal shear assumes compression struts form in the concrete slab

6.2.4 (4)

Verify that:

$v_{Ed} \le v f_{cd} \sin \theta_f \cos \theta_f$

BS EN 1992-1-1 NA 2 Table NA.1

where:

$$v = 0.6 \left[1 - \frac{f_{ck}}{250} \right]$$

$$v = 0.6 \left[1 - \frac{25}{250} \right] = 0.54$$

$v f_{cd} \sin \theta_f \cos \theta_f = 0.54 \times 14.2 \times \sin 26.5 \times \cos 26.5 = 3.06 \text{ N/mm}^2$

$v_{Ed} = 1.58 \text{ N/mm}^2 < 3.06 \text{ N/mm}^2$

Therefore the crushing resistance of the compression strut is adequate.

3) If the contribution of decking is included, the transverse reinforcement provided can be reduced.

| Example 03 Simply supported composite secondary beam | Sheet | 10 | of | 10 | Rev |

Serviceability limit state

Performance at the serviceability limit state should be verified. However, no verification is included here. The National Annex for the country where the building is to be constructed should be consulted for guidance.

Considerations would be:

- Short-term, long-term and dynamic modular ratios

- Serviceability combinations of actions

- Composite bending stiffness of the beam

- Total deflection and deflection due to imposed loads

- Stresses in steel and concrete (to validate deflection assumptions)

- Natural frequency.

| Example 03 Simply supported composite secondary beam | Sheet | 10 | of | 10 | Rev |

Serviceability limit state

	Job No.			Sheet	1	of 6	Rev	C

Job Title	Example no. 04	Revised by DGB, April 09
Subject	Edge beam	

Silwood Park, Ascot, Berks SL5 7QN
Telephone: (01344) 636525
Fax: (01344) 636570

CALCULATION SHEET

Client		Made by	MXT	Date	Dec 2007
		Checked by	AB	Date	April 2007

Unless stated otherwise all references are to BS EN 1993-1-1

Edge beam with torsion

Each 6 m span edge beams is unrestrained along its length. It carries permanent loads only. The brickwork load is applied with an eccentricity of 172 mm to the centroidal axis and induces torsion to the beam. The chosen member is a RHS, which is excellent at resisting torsion.

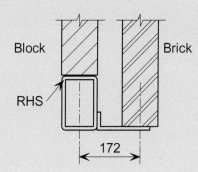

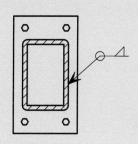

End detail

Actions

Permanent actions

Uniformly Distributed Load (brickwork)	g_1	= 4.8 kN/m
Uniformly Distributed Load (blockwork)	g_2	= 3.0 kN/m
Uniformly Distributed Load (assumed self weight)	g_3	= 0.47 kN/m

Ultimate Limit State (ULS)

Partial factors for actions

BS EN 1990 A1.3.1(4)

For the design of structural members not involving geotechnical actions, the partial factors for actions to be used for ultimate limit state design should be obtained from Table A1.2(B) and the National Annex

NA 2.2.3.2

Table NA.A1.2(B)

Partial factor for permanent actions	γ_G	= 1.35
Reduction factor	ζ	= 0.925

Combination of actions for ULS

This example uses BS EN 1990 Equation 6.10b. Expression 6.10a should also be checked, which may be more onerous.

BS EN 1990
Table A1.2(B)
& Eq. (6.10b)

UDL (total permanent)

$$F_d = \xi \times \gamma_G (g_1 + g_2 + g_3) \text{ kN/m}$$

$$F_d = 0.925 \times 1.35 \times (4.8 + 3.0 + 0.47) = 10.33 \text{ kN/m}$$

Example 04 Edge beam | Sheet 2 of 6 | Rev

UDL (permanent, inducing torsion)

$$F_{d,T} = \xi \times \gamma_G G_1 = 0.925 \times 1.35 \times 4.8 = 5.99 \text{ kN/m}$$

Design moments and shear force

Span of beam $\quad\quad\quad\quad\quad\quad\quad L \quad = 6000 \text{ mm}$
Eccentricity of brickwork $\quad\quad\quad\quad e \quad = 172 \text{ mm}$
Maximum design bending moment occurs at the mid-span

$$M_{Ed} = \frac{F_d L^2}{8} = \frac{10.33 \times 6^2}{8} = 46.5 \text{ kNm}$$

Design bending moment
$M_{Ed} = 46.5 \text{kNm}$

Maximum design shear force occurs at the supports

$$V_{Ed} = \frac{F_d L}{2} = \frac{10.33 \times 6}{2} = 31.0 \text{ kN}$$

Design shear force
$V_{Ed} = 31.0 \text{ kN}$

Maximum design torsional moment occurs at the supports

$$T_{Ed} = \frac{F_{Ed,T} \times e \times L}{2} = \frac{5.99 \times 0.172 \times 6}{2} = 3.1 \text{ kNm}$$

Design torsional moment
$T_{Ed} = 3.1 \text{ kNm}$

The design bending moment, torsional moment and shear force diagrams are shown below.

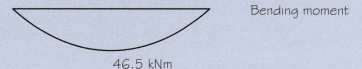

Bending moment

46.5 kNm

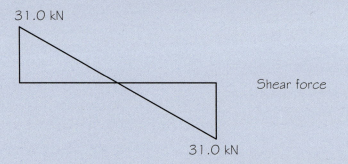

31.0 kN

Shear force

31.0 kN

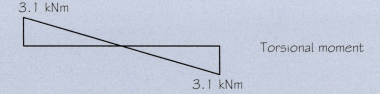

3.1 kNm

Torsional moment

3.1 kNm

| Example 04 Edge beam | Sheet | 3 | of | 6 | Rev |

Trial section

P363

Try 250 × 150 × 8.0 RHS in S355 steel. The RHS is class 1 under the given loading.

Depth of section	h	= 250 mm
Width of section	b	= 150 mm
Wall thickness	t	= 8 mm
Plastic modulus about the y-axis	$W_{pl,y}$	= 501 cm³
Cross-sectional area	A	= 6080 cm²
St Venant torsional constant	I_T	= 5020 cm⁴
Torsional section modulus	W_t	= 506 cm³
Second moment of area about z-z axis	I_z	= 2300 cm⁴

NA 2.4

BS EN
10210-1
Table A3

For steel grade S355 and $t < 16$ mm

Yield strength f_y = 355 N/mm²

Partial factors for resistance

NA 2.15

γ_{MO} = 1.0

Resistance of the cross section

Note that the following verification assumes that the maximum shear, bending and torsion are coincident, which is conservative.

6.2.6

Plastic shear resistance

$$V_{pl,Rd} = \frac{A_v \left(f_y / \sqrt{3} \right)}{\gamma_{MO}}$$

Where $A_v = \dfrac{Ah}{(b+h)} = \dfrac{6080 \times 250}{(250+150)} = 3800$ mm²

$$V_{pl,Rd} = \frac{3800 \left(355 / \sqrt{3} \right)}{1.0 \times 10^3} = 779 \text{ kN}, \; > 28.5 \text{ kN, OK}$$

6.2.6(6)

Shear buckling resistance

The shear buckling resistance for webs should be checked according to section 5 of BS EN 1993-1-5 if:

Eq. 6.22

$$\frac{h_w}{t_w} > \frac{72\varepsilon}{\eta}$$

Table 5.2

$$\varepsilon = \sqrt{\frac{235}{f_y}} = \sqrt{\frac{235}{355}} = 0.81$$

| Example 04 Edge beam | Sheet | 4 | of | 6 | Rev |

6.2.6(6)	$\eta = 1.0$ (conservative)

$h_w = h - 3t = 250 - 3 \times 8.0 = 226$ mm

$$\frac{h_w}{t_w} = \frac{226}{8.0} = 28.3$$

$$\frac{72\varepsilon}{\eta} = \frac{72 \times 0.81}{1.0} = 58$$

28.3 < 58 Therefore the shear buckling resistance of the web does not need to be checked.

6.2.7   *Torsional resistance*

The torsional moment may be considered as the sum of two internal effects:

$T_{Ed} = T_{t,Ed} + T_{w,Ed}$

6.2.7(7)   But $T_{w,Ed}$ may be neglected for hollow sections

For a closed section, $T_{Rd} = \dfrac{f_y W_t}{\sqrt{3} \times \gamma_{MO}}$

$$= \frac{355 \times 506 \times 10^3}{\sqrt{3} \times 1.0} \times 10^{-6} = 103.7 \text{ kNm}$$

103.7 > 3.1, OK

6.2.7(9)   *Shear and torsion*

Eqn 6.25 $\dfrac{V_{Ed}}{V_{pl,T,Rd}} \leq 1.0$

For a structural hollow section

$$V_{pl,T,Rd} = \left[1 - \frac{\tau_{t,Ed}}{\left(f_y / \sqrt{3}\right)/\gamma_{MO}} \right] \times V_{pl,Rd}$$

Shear stress due to torsion, $\tau_{t,Ed} = \dfrac{T_{t,Ed}}{W_t}$

$$\tau_{t,Ed} = \frac{3.1 \times 10^6}{506 \times 10^3} = 6.1 \text{ N/mm}^2$$

Then $V_{pl,T,Rd} = \left[1 - \dfrac{\tau_{t,Ed}}{\left(f_y / \sqrt{3}\right)/\gamma_{MO}} \right] \times V_{pl,Rd}$

$$V_{pl,T,Rd} = \left[1 - \frac{5.5}{\left(355/\sqrt{3}\right)/1.0} \right] \times 779 = 758 \text{ kN}$$

31.0 < 758, OK

6.2.8(2) **_Bending and shear_**

The shear force (V_{Ed} = 31.0 kN) is less than half the plastic shear resistance ($V_{pl,Rd}$ = 779 kN), so no reduction in the bending resistance due to the presence of shear is required.

6.2.8(4) **_Bending, shear, and torsion_**

The shear force (V_{Ed} = 31.0 kN) is less than half the plastic shear resistance accounting for torsional effects ($V_{pl,T,Rd}$ = 758 kN), so ρ = 0 and therefore the yield strength used in calculating the bending resistance need not be reduced.

Bending resistance

6.2.5 **_Cross section resistance_**

6.2.5(2) The design resistance for bending for Class 1 and 2 cross-sections is:

$$M_{c,Rd} = M_{pl,Rd} = \frac{W_{pl,y} f_y}{\gamma_{M0}} = \frac{501 \times 10^3 \times 355}{1.0 \times 10^6} = 177.9 \text{ kNm}$$

Design bending resistance, $M_{c,Rd}$ = 177.9 kNm

177.9 > 46.5, OK

6.3.2 **_Buckling resistance_**

6.3.2.2(4) For slendernesses $\bar{\lambda}_{LT} < \bar{\lambda}_{LT,0}$ lateral torsional buckling effects may be ignored.

NA 2.17 $\bar{\lambda}_{LT,0} = 0.4$

6.3.2.2(1) The slenderness $\bar{\lambda}_{LT}$ is given by $\bar{\lambda}_{LT} = \sqrt{\frac{W_y \times f_y}{M_{cr}}}$

Access-steel document SN003a-EN-EU

For non-destabilising loads, and where warping is neglected, the elastic critical moment for lateral-torsional buckling, M_{cr} is given by:

$$M_{cr} = C_1 \frac{\pi^2 EI}{L^2} \left\{ \sqrt{\frac{L^2 G I_T}{\pi^2 E I_z}} \right\}$$

Where:

E	is the modulus of elasticity (E = 210000 N/mm^2)
G	is the shear modulus (G = 81000 N/mm^2)
I_z	is the second moment of area about the minor axis
I_T	is the St Venant torsional constant
L	is the beam length between points of lateral restraint
C_1	accounts for actual moment distribution
C_1	= 1.127 (for simply supported beam with a UDL).

Access-steel document SN003a Table 3.2

| Example 04 Edge beam | Sheet | 6 | of | 6 | Rev |

$$M_{cr} = 1.127 \times \frac{\pi^2 \times 210000 \times 2300 \times 10^4}{6000^2}$$

$$\times \left\{ \sqrt{\frac{6000^2 \times 81000 \times 5020 \times 10^4}{\pi^2 \times 210000 \times 2300 \times 10^4}} \right\}$$

Hence, $M_{cr} = 2615$ kNm

6.3.2.2(1)

$$\overline{\lambda}_{LT} = \sqrt{\frac{W_y \times f_y}{M_{cr}}} = \sqrt{\frac{W_{pl,y} \times f_y}{M_{cr}}} = \sqrt{\frac{501 \times 10^3 \times 355}{2615 \times 10^6}} = 0.26$$

0.26 < 0.4, so lateral-torsional buckling effects can be neglected.

Serviceability limit state (SLS)

Twist at SLS

In this case, twist will be assessed under the unfactored permanent loads.

Partial factors for actions

Partial factor for permanent actions $\qquad \gamma_G = 1.0$

Maximum torsional moment $= \dfrac{3.1 \times 1.0}{1.35 \times 0.925} = 2.48$ kNm

Maximum twist per unit length is given by:

$$\text{Twist} = \frac{T_{Ed}}{GI_t} = \frac{2.48 \times 10^6}{81000 \times 5020 \times 10^4}$$

$= 6.1 \times 10^{-7}$ radians/mm

Twist at midspan $= 0.5 \times 6.1 \times 10^{-7} \times 3000 = 0.9 \times 10^{-3}$ radians

$= 0.05$ degrees

Note that this calculation assumes that the support conditions prevent any form of twisting – so friction grip connections or similar may be required.

Right margin:

$M_{cr} = 2615$ kNm

$\overline{\lambda}_{LT} = 0.26$

	Job No.		Sheet	*1*	of *4*	Rev	C
	Job Title	Example no. 05	Revised by DGB, April 09				
	Subject	Column in Simple Construction					
Silwood Park, Ascot, Berks SL5 7QN	Client		Made by	LG		Date	Dec 2007
Telephone: (01344) 636525 Fax: (01344) 636570							
CALCULATION SHEET			Checked by	PS		Date	Dec 2007

Unless stated otherwise all references are to BS EN 1993-1-1:2005	**Column in Simple Construction** *Description* This example demonstrates the design of an internal column in simple construction. Note that the internal columns do not carry roof loads. *Internal column at ground level – Gridline G2* Column height $= 5.0$ m **Actions**	
See structural arrangement and actions	Reactions at each of the three floor levels from 8 m span beams: Permanent $\quad = 0.5 \times 8 \times 6 \times 3.7 = 88.8$ kN Variable $\quad = 0.5 \times 8 \times 6 \times 3.3 = 79.2$ kN Reactions at each of the three floor levels from 6 m span beams: Permanent $\quad = 0.5 \times 6 \times 6 \times 3.7 = 66.6$ kN Variable $\quad = 0.5 \times 6 \times 6 \times 3.3 = 59.4$ kN The total load acting on the column due to three floors is given by: Permanent $\quad G_k = 3 \times (88.8 + 66.6) = 466.2$ kN Variable $\quad Q_k = 3 \times (79.2 + 59.4) = 415.8$ kN **Ultimate Limit State (ULS)**	$G_k = 466.2$ kN $Q_k = 415.8$ kN
BS EN 1990 Table A1(2)B NA 2.2.3.2 Table NA.A1.2(B)	*Partial factors for actions* For permanent actions $\qquad \gamma_G = 1.35$ For variable actions $\qquad \gamma_Q = 1.5$ *Reduction factor* $\xi = 0.925$	
BS EN 1990 6.4.3.2	*Design value of combined actions, from equation 6.10b* $= \xi \gamma_G \, G_k + \gamma_Q \, Q_k$ $= 0.925 \times 1.35 \times 466.2 + 1.5 \times 415.8 = 1206$ kN At level 1, The reaction from an 8 m beam is $0.925 \times 1.35 \times 88.8 + 1.5 \times 79.2 = 230$ kN The reaction from an 6 m beam is $0.925 \times 1.35 \times 66.6 + 1.5 \times 59.4 = 172$ kN	The ULS axial load, $N_{Ed} = 1206$ kN

| Example 05 Column in Simple Construction | Sheet | 2 | of | 4 | Rev |

Partial factors for resistance

6.1(1)

NA 2.15

γ_{M0} = 1.0

γ_{M1} = 1.0

Trial section

Try 254 × 254 × 73 UKC, S275

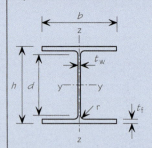

SCI P363

Depth	h	= 254.1 mm
Width of cross-section	b	= 254.6 mm
Flange thickness	t_f	= 14.2 mm
Web thickness	t_w	= 8.6 mm
Radius of gyration	i_z	= 6.48 cm
Section area	A	= 93.1 cm²
Plastic modulus, y-y	$W_{pl,y}$	= 992 cm³

NA 2.4

BS EN
10025-2

Table 7

Yield Strength, f_y

Steel grade = S275

Nominal thickness of the element, $t \le 16$ mm then f_y = 275 N/mm²

P363

Section classification

Cross-section is assumed to be at Class 1 or 2.
(no UKC is Class 4 under compression alone; only a 152 UKC 23 is not class 2 or better under bending alone in S275)

Access Steel
document
SN008a-EN-
EU

Buckling lengths

Buckling length about y-y axis $L_{cr,y}$ = 5.0 m

Buckling length about z-z axis $L_{cr,z}$ = 5.0 m

Design moments on column due to beam reactions

Access Steel
document
SN005a-EN-
EU

For columns in simple construction the beam reactions are assumed to act at 100 mm from the face of the column.

In the minor axis, the beam reactions at internal columns are identical and hence there are no minor axis moments to be considered.

| Example 05 Column in Simple Construction | Sheet *3* | of *4* | Rev |

Reactions at level 1, for major axis bending

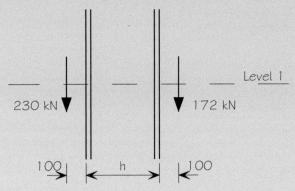

Level 1

230 kN   172 kN

100 |← h →| 100

$M_{1,y,Ed} = ((h/2) + 100) \times (230 - 172) = 13.2$ kNm

The moment is distributed between the column lengths above and below level 1 in proportion to their bending stiffness (*EI/L*), unless the ratio of the stiffnesses does not exceed 1.5 – in which case the moment is divided equally. As the ratio of the column stiffnesses is less than 1.5, the design moment at level 1 is therefore:

$M_{y,Ed} = 13.2 \times 0.5 = 6.6$ kNm $M_{y,Ed} = 6.6$ kNm

$M_{z,Ed} =$ zero because the beam reactions are equal on either side of the column in this direction as the bay widths and loads are identical. $M_{z,Ed} = 0$

Flexural buckling resistance

6.3.1.3

$\lambda_1 = 93.9\varepsilon = 93.9 \times (235/275)^{0.5} = 86.8$

$\bar{\lambda}_z = \dfrac{L_{cr}/i_z}{\lambda_1} = \dfrac{5000/64.8}{86.8} = 0.89$

Table 6.2

$h/b < 1.2$ and $t_f < 100$ mm, so use buckling curve 'c' for the z-axis for flexural buckling.

Figure 6.4

From graph, $\chi_z = 0.61$

Eq. (6.47)

$N_{b,z,Rd} = \chi_z Af/\gamma_{M1} = 0.61 \times 9310 \times 275 \times 10^{-3}/1.0 = 1562$ kN $N_{b,z,Rd} = 1562$ kN

6.3.2

Lateral torsional buckling resistance moment

Conservatively the slenderness for lateral torsional buckling may be determined as:

Access Steel document SN002a-EN-EU

$\bar{\lambda}_{LT} = 0.9\bar{\lambda}_z = 0.9 \times 0.89 = 0.80$

(Other methods for determining $\bar{\lambda}_{LT}$ may provide a less conservative design, as illustrated in example 02.)

6.3.2.3

For rolled and equivalent welded sections

$\chi_{LT} = \dfrac{1}{\phi_{LT} + \sqrt{\phi_{LT}^2 - \beta\bar{\lambda}_{LT}^2}}$

where $\phi_{LT} = 0.5\left[1 + \alpha_{LT}\left(\bar{\lambda}_{LT} - \bar{\lambda}_{LT,0}\right) + \beta\bar{\lambda}_{LT}^2\right]$

| Example 05 Column in Simple Construction | Sheet | 4 | of | 4 | Rev |

6.3.2.3	$\bar{\lambda}_{LT,0}$ = 0.4
NA 2.17	β = 0.75
Table 6.5	For rolled bi-symmetric I-sections with $h/b \leq 2$: use buckling curve 'b'.
Table 6.3	For buckling curve 'b', α_{LT} = 0.34

$$\phi_{LT} = 0.5\left(1 + 0.34(0.80 - 0.40) + 0.75 \times 0.80^2\right) = 0.81$$

$$\chi_{LT} = \frac{1}{0.81 + \sqrt{0.81^2 - 0.75 \times 0.80^2}} = 0.81$$

6.3.2.3

But the following restrictions apply:

$$\chi_{LT} \leq 1.0$$

$$\chi_{LT} \leq \frac{1}{\bar{\lambda}_{LT}^2} = \frac{1}{0.80^2} = 1.56$$

$$\therefore \ \chi_{LT} = 0.81$$

6.3.2.1(3)

$$M_{b,Rd} = \frac{\chi_{LT} W_y f_y}{\gamma_{M1}} = \frac{\chi_{LT} W_{pl,y} f_y}{\gamma_{M1}} \text{ for Class 1 or 2 cross-sections}$$

$$= \frac{0.81 \times 992 \times 275 \times 10^{-3}}{1.0} = 221 \text{ kNm}$$

$M_{b,Rd}$ = 221 kNm

SN048a-EN-GB Access Steel document

Combined bending and axial compression buckling (simplified)

Instead of equation 6.61 and 6.62, the simplified expression given below is used:

$$\frac{N_{Ed}}{N_{b,z,Rd}} + \frac{M_{y,Ed}}{M_{b,Rd}} + 1.5\frac{M_{z,Ed}}{M_{z,Rd}} \leq 1.0$$

$$\frac{1206}{1562} + \frac{6.6}{221} + 0 = 0.80 \leq 1.0$$

Therefore a 254 × 254 × 73 UKC is adequate.

Section used is 254×254×73 UKC, S275

BS EN 1991-1-1 Tables 6.9 & 6.10 NA 2.10 Table NA 7 Unless stated otherwise all references are to BS EN 1993-1-1	## Roof Truss

The truss to be designed is to support a roof which is only accessible for normal maintenance and repair. The truss is 14 m span with 15° pitch. The dimensions of the truss are shown in the figure below. The imposed roof load due to snow obtained from BS EN 1991-1-3 is less than 0.6 kN/m², therefore the characteristic imposed roof load is taken from BS EN 1991-1-1 and the National Annex. The truss uses hollow sections for its tension chord, rafters, and internal members. The truss is fully welded. Truss analysis is carried out by placing concentrated loads at the joints of the truss. All of the joints are assumed to be pinned in the analysis and therefore only axial forces are carried by members. |

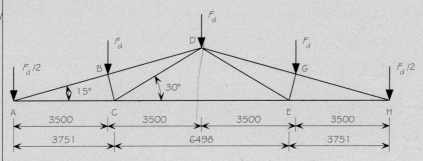

Characteristic actions

Permanent actions

Self weight of roof construction	0.75 kN/m²
Self weight of services	0.15 kN/m²
Total permanent actions	0.90 kN/m²

Variable actions

Imposed roof load	0.60 kN/m²
Total imposed action	0.60 kN/m²

Ultimate Limit State (ULS)

NA 2.2.3.2 Table NA.A1.2(B)

Partial factors for actions

Partial factor for permanent actions $\gamma_G = 1.35$

Partial factor for variable actions $\gamma_G = 1.5$

Reduction factor $\xi = 0.925$

Design value of combined actions, using equation 6.10b

$= 0.925 \times 1.35 \times 0.9 + 1.5 \times 0.6 = 2.02 \text{ kN/m}^2$

| Example 06 Roof truss | Sheet 2 of 5 | Rev |

Design values of combined actions on purlins supported by truss

For the distance of 3.5 m between purlins centre to centre

Design value $= 2.02 \times 3.5/\cos 15° = 7.32$ kN/m

Design value of combined actions on truss

For a purlin span of 6 m

$F_d = 7.32 \times 6 = 43.92$ kN

$F_d = 43.92$ kN

Truss analysis (due to forces F_d)

Reaction force at support A $R_A = 2 \times F_d = 87.8$ kN

At joint A	$F_{AB} \times \sin 15° + (R_A - W/2) = 0$	$F_{AB} = -255$ kN
	$F_{AB} \times \cos 15° + F_{AC} = 0$	$F_{AC} = 246$ kN
At joint B	$F_{BC} + W \times \cos 15° = 0$	$F_{BC} = -42$ kN
	$F_{BD} - F_{AB} - W \times \sin 15° = 0$	$F_{BD} = -243$ kN
At joint C	$F_{BC} \times \sin 75° + F_{CD} \times \sin 30° = 0$	$F_{CD} = 82$ kN
	$F_{CE} - F_{AC} - F_{BC} \times \cos 75° + F_{CD} \times \cos 30° = 0$	$F_{CE} = 164$ kN

Partial factors for resistance

6.1(1)
NA 2.15

$\gamma_{M0} = 1.0$

$\gamma_{M1} = 1.0$

$\gamma_{M2} = 1.25$

Design of Top Chords (members AB, BD, DG, GH)

$N_{Ed} = 255$ kN

Maximum design force (member AB and GH) = 255 kN (compression)

Try $100 \times 100 \times 5$ square hollow section in S355 steel

NA 2.4

BS EN
10210-1

Table A3

Material properties:

modulus of elasticity $E = 210000$ N/mm^2

steel grade S355 and thickness ≤ 16 mm

Yield strength $f_y = 355$ N/mm^2

$\varepsilon = \sqrt{\dfrac{235}{f_y}} = \sqrt{\dfrac{235}{355}} = 0.81$

P363

Section properties:

Depth and width of section $h, b = 100$ mm

Thickness $t = 5$ mm

Radius of gyration $i_z = 38.6$ mm

Area $A = 1870$ mm^2

56

bibliotheca SelfCheck System

Customer ID: B0007384998

Items that you have checked out

Title:
 Steel building design : worked examples for
 students : in accordance with Eurocodes and
 the UK National Annexes
ID: 6546460802
Due: 05 February 2019

Title:
 Design of structural elements [electronic
 resource] : concrete, steelwork, masonry and
 timber designs to British standards and
 Eurocodes
ID: 054024073X
Due: 19 February 2019

Total items: 2
Account balance: £0.00
22/01/2019 15.43
Checked out: 5
Overdue: 0
Hold requests: 0
Ready for collection: 0

Thank you for using the bibliotheca SelfCheck
System.

Table 5.2	*Classification of the cross-section:*	
	$c = 100 - 3 \times 5 = 85$ mm	
	$\dfrac{c}{t} = \dfrac{85}{5} = 17$	
	Class 3 limit $= 42e = 42 \times 0.81 = 34.$	The section is at least Class 3
	$17 < 34$, so the section is at least class 3	
Eq.(6.10) for Class 3 sections	**Compression resistance of the cross-section:**	
	$N_{c,Rd} = \dfrac{Af_y}{\gamma_{MO}} = \dfrac{1870 \times 355 \times 10^3}{1.0} = 663$ kN	
	$\dfrac{N_{Ed}}{N_{c,Rd}} = \dfrac{255}{633} = 0.40 < 1.0$	
	Therefore, the compressive design resistance is adequate.	$N_{c,Rd} > N_{Ed}$
Eq.(6.50) for Class 1,2 and 3 cross-sections	**Flexural buckling resistance:**	
	Determine the non-dimensional slenderness for flexural buckling:	
	$\bar{\lambda}_z = \sqrt{\dfrac{Af_y}{N_{cr}}} = \dfrac{L_{cr}}{i_z}\dfrac{1}{\lambda_1}$	
	where $L_{cr} = 1.0 \times L_{AB} = \dfrac{3500}{\cos 15^\circ} = 3623$ mm	
	$\lambda_1 = \pi\sqrt{\dfrac{E}{f_y}} = \pi\sqrt{\dfrac{210000}{355}} = 76.4$	
	$\bar{\lambda} = \sqrt{\dfrac{Af_y}{N_{cr}}} = \dfrac{L_{cr}}{i_z}\dfrac{1}{\lambda_1} = \dfrac{3623}{38.6}\dfrac{1}{76.4} = 1.23$	
Eq.(6.49) and Tables 6.1 and 6.2	Determine the reduction factor due to buckling	
	$\chi = \dfrac{1}{\Phi + \sqrt{\Phi^2 - \bar{\lambda}^2}}$	
	where: $\Phi = 0.5\left[1 + \alpha(\bar{\lambda} - 0.2) + \bar{\lambda}^2\right]$	
	$\alpha = 0.21$ (use buckling curve 'a' for a SHS)	
	$\Phi = 0.5\left[1 + 0.21(1.23 - 0.2) + 1.23^2\right] = 1.36$	
	$\chi_z = \dfrac{1}{\Phi + \sqrt{\Phi^2 - \bar{\lambda}^2}} = \dfrac{1}{1.36 + \sqrt{1.36^2 - 1.23^2}} = 0.52$	
	$N_{b,Rd} = \dfrac{\chi_z Af_y}{\gamma_{M1}} = \dfrac{0.52 \times 1870 \times 355 \times 10^{-3}}{1.0} = 345$ kN	

| Example 06 Roof truss | | Sheet **4** | of **5** | Rev |

$$\frac{N_{Ed}}{N_{b,Rd}} = \frac{255}{345} = 0.71 \ < \ 1.0, \ OK$$

Therefore, the design flexural buckling resistance of the selected 100 × 100 × 5 SHS is satisfactory.

$N_{b,Rd} > N_{Ed}$

Design of bottom chords (members AC, CE, EH)

Maximum design force (member AC and EH) = 246 kN (in tension)

N_{Ed} = 246 kN

The bottom chord will also be a 100 × 100 × 5 SHS, S355. By inspection, the design tension resistance is equal to the design plastic resistance of the cross section.

Eq.(6.6)

$$N_{pl,Rd} = \frac{Af_y}{\gamma_{M0}} = \frac{1870 \times 355 \times 10^{-3}}{1.0} = 663 \ kN$$

663 kN > 246 kN, OK

$N_{pl,Rd} > N_{Ed}$

Design of internal members (members BC, EG, CD, DE)

Maximum design compression force (BC and EG) = 42 kN
Maximum design tension force (CD and DE) = 82 kN
Maximum length in compression is BC and EG = 970 mm

N_{Ed} = 42 kN

Try a 70 × 70 × 5 SHS, in S355 steel.

Following the same design process as above, the following resistances can be calculated:

Flexural buckling resistance (L_{cr} = 970mm), $N_{b,Rd}$ = 419 kN

Tension resistance, $N_{pl,Rd}$ = 450 kN

Thus all internal members will be selected as 70 × 70 × 5 SHS, in S355 steel.

BS EN 1993-1-1

NA 2.23

Serviceability limit state (SLS)

The UK National Annex provides suggested limits for vertical and horizontal deflections. The National Annex also states that the deflections should be checked under unfactored variable loads and that permanent loads should not be included.

Partial factors for actions

Partial factor for variable actions γ_G = 1.0

Design value of combined actions

= 1.0 × 0.6 = 0.6 kN/m^2

Design value of combined actions on truss

= 6 × 0.6 × 3.5/Cos 15° = 13.0 kN

F_d = 13.0 kN

| Example 06 Roof truss | Sheet 5 | of 5 | Rev |

Deflection

The maximum allowable deflection is assumed to be span/300;

Span/300 = 14000/300 = 46.67 mm.

The maximum deflection of the truss is obtained for the SLS value of combined actions (i.e. F_d = 13.4 kN). The deflection at the apex was found to 6.4 mm when all of the joints are assumed to be pinned. Deflection is therefore satisfactory.

Connections

The design of the connections is not shown in this example, although this is particularly important for trusses fabricated from hollow sections. The joint resistances depend on the type of joint, the geometry of the joint and the forces in the members. It is unlikely that the joints in hollow section fabrications can carry as much load as the members themselves, without expensive strengthening, which should be avoided.

Joint resistance should be checked at the design stage, so that appropriate members can be chosen to ensure that in addition to the members resisting the design load, the joints can also transfer the member forces without strengthening.

The design of hollow section joints is covered in BS EN 1993-1-8

<table>
<tr><td rowspan="4">
SCI

Silwood Park, Ascot, Berks SL5 7QN
Telephone: (01344) 636525
Fax: (01344) 636570

CALCULATION SHEET</td><td>Job No.</td><td colspan="2">Sheet   *1*   of *2*   Rev   C</td></tr>
<tr><td>Job Title   Example no. 07</td><td colspan="2">Revised by DGB, April 09</td></tr>
<tr><td>Subject   Choosing a steel sub-grade</td><td colspan="2"></td></tr>
<tr><td>Client</td><td>Made by   LPN   Date   May 2007</td><td></td></tr>
</table>

	Checked by MEB Date Jan 2008

Unless stated otherwise all references are to BS EN 1993-1-10

Choosing a steel sub-grade

Introduction

Determine the steel sub-grade that may be used for the simply supported restrained beam (UKB 457 × 191 × 82 steel grade S275).

The example follows the procedure recommended in PD 6695-1-10. This published document provides non conflicting complimentary information to the Eurocode, and presents a straightforward approach to the choice of steel sub-grade.

Floor beam at Level 1 – Gridline G1-2

Beam span,	L =	8.0m
Bay width,	w =	6.0m

Actions

Permanent action :	g_k =	3.7 kN/m^2
Variable action :	q_k =	3.8 kN/m^2

SCI P363

Section Properties

From example 01:

Web thickness	t_w =	9.9 mm
Flange thickness	t_f =	16.0 mm
Elastic modulus, y-y	$W_{el,y}$ =	1610.9 cm^3
Yield strength	f_y =	275 N/mm^2

Combination of actions

2.2.4 (i)

Effects are combined according to the following expression:

BS EN 1990
A.1.2.2 (1)

$$E_d = E\{ A[T_{Ed}] \;"+"\; \Sigma G_K \;"+"\; \psi_1 \, Q_{K1} \;"+"\; \underline{\Sigma \, \psi_{2,i} \, Q_{Ki}} \}$$

NA 2.2.2
Table
NA.A1.1

not relevant for this example as there is only one variable action

where ψ_1 = 0.5 (Category B: Office areas)

It is assumed that there are no locked in stresses due to temperature, since bolts in clearance holes are used. Therefore $A[T_{Ed}]$ = 0

Design value of combined actions

$G_{K1} + \psi_1 \, Q_K$ = 0.5 × 3.8 × 6 = 11.4 kN/m

| Example 07 Choosing a steel sub-grade | Sheet | 2 | of | 2 | Rev |

Maximum moment at mid span :

$$M_{y,Ed} = 11.4 \times 8^2 / 8 = 91.2 \text{ kNm}$$

Design moment diagram

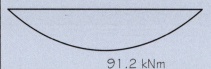

Bending moment

91.2 kNm

Calculation of maximum bending stress:

$$\sigma_{Ed} = \frac{M_{y,Ed}}{W_{el,y}} = \frac{91.2 \times 1000}{1610.9} = 56.6 \text{ N/mm}^2$$

2.3.2

Stress level as a proportion of nominal yield strength

$f_y(t)$ may be taken as the minimum yield strength from the product Standard, and is in this example.

BS EN
10025-2
Table 7

The flange is 16 mm thick

$f_y(t) = 275 \text{ N/mm}^2$

$$\sigma_{Ed} = \frac{56.6}{275} f_y(t) = 0.20 \, f_y(t)$$

For steel internal steelwork in buildings, the limiting thickness is taken from Table 2 of PD 6695-1-10

It is assumed that the Detail type is "welded – moderate"

Conservatively choosing "Comb.6" ($\sigma_{Ed}/f_y(t) = 0.3$)

PD 6695-1-
10

Table 2

S275JR gives a limiting thickness of 50 mm, > 16 mm, OK.

Steel sub-grade
S275JR is
adequate.

Job No.			Sheet *1* of *7*		Rev C
Job Title	Example no. 08		Revised by DGB, April 09		
Subject	Composite slab				
Client		Made by	ALS	Date	Nov 2007
		Checked by	MEB	Date	Jan 2008

Silwood Park, Ascot, Berks SL5 7QN
Telephone: (01344) 636525
Fax: (01344) 636570

CALCULATION SHEET

Unless stated otherwise all references are to BS EN 1994-1-1

Composite slab

Introduction

This example demonstrates the design of the composite floor slab on the second storey that is supported by the composite beam designed in Example 3. The profiled metal deck is CF60 and the thickness of the slab is 130 mm.

Verification is needed for both the construction stage (non-composite) and constructed stage (composite)[1]. Although generally checks at the non-composite stage are based on two continuous spans, for simplicity only a single span case will be considered here.

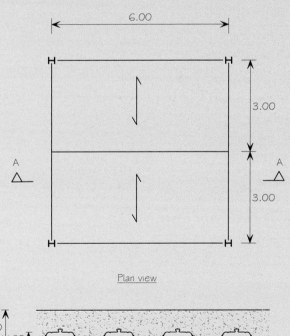

Plan view

Section A - A

The continuous floor slab will be designed as a series of simply supported spans. This approach is conservative because it does not take into account the positive effect of the continuity over the supports.

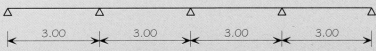

---

1) The floor slab should be designed for both the construction stage and the composite stage. During the construction stage, the metal decking acts as formwork and has to support its own weight, wet concrete, and construction loads. The resistance of the metal decking during the construction stage needs to be verified at the ultimate and serviceability limit state

| Example 08 Composite slab | Sheet *2* of *7* | Rev |

Floor slab and material properties

Total depth of slab $h = 130$ mm

Corus profiled steel sheeting CF60

Thickness of profile $t = 1.0$ mm
Depth of profile $h_p = 60$ mm
Span $L = 3$ m
Effective cross-sectional area of the profile $A_{pe} = 1424$ mm^2/m
Second moment of area of the profile $I_p = 106.15$ cm^4/m
Yield strength of the profiled deck $f_{yp} = 350$ N/mm^2
from manufacturer's data:
Design value of bending resistance (sagging)

$M_{Rd} = 11.27$ KNm/m

Height of neutral axis above soffit: $= 30.5$ mm

BS EN 1992-1-1 Table 3.1
BS EN 1991-1-1 Table A.1

Concrete

Normal concrete strength class C25/30

Density (normal weight, reinforced)   26 kN/m³ (wet)
                                       25 kN/m³ (dry)

[These density values may vary for a specific project depending on the amount of steel reinforcement.]

Cylinder strength $f_{ck} = 25$ N/mm^2
Modulus of elasticity $E_{cm} = 31$ kN/mm^2

Actions

Concrete weight

Self weight of the concrete slab (volume from decking manufacturer's data)

$0.097 \times 26 \times 10^{-6} = 2.52$ kN/m^2 (wet)

$0.097 \times 25 \times 10^{-6} = 2.43$ kN/m^2 (dry)

Permanent Actions

Construction stage	kN/m²	Composite stage	kN/m²
Steel deck	0.11	Concrete slab	2.43
Total	0.11	Steel deck	0.11
		Ceiling and services	0.15
		Total	2.69

Construction stage:
$g_k = 0.11$ kN/m^2

Composite stage:
$g_k = 2.69$ kN/m^2

Variable actions

At the construction stage, the loading considered is a 0.75 kN/m^2 load across the entire slab, with an additional 0.75 kN/m^2 load across a 3 m span, which can be positioned anywhere on the slab span. In this case the span is 3 m, and so the construction loading across the whole span is 1.50 kN/m²

Example 08 Composite slab	Sheet	3	of	7	Rev

BS EN 1991-1-6 NA 2.13	**Construction stage** kN/m² **Composite stage** kN/m² Construction loading Imposed floor load 3.30 (1) Outside the (See structural arrangement and working area 0.75 loading) (2) Inside the working area (additional) 0.75 (3) Concrete slab 2.52 Total 4.02	Construction stage: q_k = 4.02 kN/m² Composite stage: q_k = 3.30 kN/m²

<u>Ultimate Limit State (ULS)</u>

BS EN 1990
NA 2.2.3.2
Table
NA.A1.2(B)

Partial factors for actions

Partial factor for permanent actions	γ_G	= 1.35
Partial factor for variable actions	γ_Q	= 1.5
Reduction factor	ξ	= 0.925

Combination of actions at ULS

The design values may be calculated using expression 6.10, or the more onerous of expression 6.10a and 6.10b. In this example, the use of expression 6.10b is demonstrated.

BS EN 1990
Eqn. 6.10b

Design value of combined actions = $\xi\gamma_G g_k + \gamma_Q q_k$

Construction stage:

Distributed load $(0.925 \times 1.35 \times 0.11) + (1.5 \times 4.02) = 6.17$ kN/m²

Construction stage
F_d = 6.17 kN/m²

Composite stage:

Distributed load $(0.925 \times 1.35 \times 2.69) + (1.5 \times 3.3) = 8.31$ kN/m²

Composite stage
F_d = 8.31 kN/m²

<u>Design moment and shear force</u>

Construction Stage

The design bending moment per metre width of the steel deck is:

$$M_{Ed} = \frac{F_d L^2}{8} = \frac{6.17 \times 3^2}{8} = 6.94 \text{ kNm/m width}$$

M_{Ed} = 6.94 kNm/m

The design shear force per metre width of the steel deck is:

$$V_{Ed} = \frac{F_d L}{2} = \frac{6.17 \times 3}{2} = 9.26 \text{ kN/m}$$

V_{Ed} = 9.26 kN/m

Normal Stage

The design bending moment per metre width of the steel deck is:

$$M_{Ed} = \frac{F_d L^2}{8} = \frac{8.31 \times 3^2}{8} = 9.35 \text{ kNm/m width}$$

M_{Ed} = 9.35 kNm/m

The design shear force per metre width of the steel deck is:

$$V_{Ed} = \frac{F_d L}{2} = \frac{8.31 \times 3}{2} = 13.07 \text{ kN/m}$$

V_{Ed} = 13.07 kN/m

| Example 08 Composite slab | Sheet | 4 | of | 7 | Rev |

BS EN 1993-1-1
NA 2.15

Partial factors for resistance

Structural steel $\qquad \gamma_{MO} = 1.0$

BS EN 1992-1-1
NA 2
Table NA.1

Concrete $\qquad \gamma_c = 1.5$
Reinforcement $\qquad \gamma_s = 1.15$

Longitudinal shear $\qquad \gamma_{VS} = 1.25$

Design values of material strengths

Steel deck

Design yield strength $\quad f_{yp,d} = \dfrac{f_{yp}}{\gamma_{MO}} = \dfrac{350}{1.0} = 350$ N/mm²

Concrete

BS EN 1992-1-1
NA 2
Table NA.1

Design value of concrete compressive strength $f_{cd} = \alpha_{cc} \times f_{ck}/\gamma_c$

$\alpha_{cc} = 0.85$

$f_{cd} = 0.85 \times 25/1.5 = 14.2$ N/mm²

$f_{cd} = 14.2$ N/mm²

BS EN 1993-1-3
6.1.1

Verification at the construction stage

Bending resistance

$\dfrac{M_{Ed}}{M_{Rd}} = \dfrac{6.94}{11.27} = 0.62 < 1.0$

Therefore the bending moment resistance at the construction stage is adequate

Shear resistance and bearing resistance

Procedures are set out in BS EN 1993-1-3 6.1.7.3. In practice, design is normally carried out by using load-span tables or by using software, which are based on testing, not calculation. Good detailing practice avoids certain failure modes.

Serviceability Limit State (SLS)

Construction Stage Deflections

9.3.2(2)

Deflection without ponding

At serviceability, loading = 0.11 + 2.52 = 2.63 kN/m²

$\delta_s = \dfrac{5 F_d L^4}{384 EI} = \dfrac{5 \times 2.63 \times 3^4}{384 \times 210 \times 106.15 \times 10} \times 10^6 = 12.4$ mm

As this is less than 10% of the slab depth (13 mm), the effects of the additional concrete may be ignored in the design of the steel sheeting.

NA 2.15

$\delta_{s,max} = L/180$ but 20mm max where the loads from ponding are ignored.

$\delta_{s,max} = 3000/180 = 16.6$ mm, OK

| Example 08 Composite slab | | | Sheet | 5 | of | 7 | Rev |

Verification of the composite slab

Ultimate Limit State(ULS)

Bending resistance – location of plastic neutral axis (pna)

6.2.1.2

Maximum compressive design force per metre in the concrete above the sheeting assuming the pna is below the solid part of the slab is determined as:

$$N_c = f_{cd} A_c = 14.2 \times 70 \times 1000 \times 10^{-3} = 994 \, kN/m$$

Maximum tensile resistance per metre of the profiled steel sheet is determined as:

$$N_p = f_{yp,d} A_p = 350 \times 1424 \times 10^{-3} = 498.4 \, kN/m$$

As $N_p < N_c$ the neutral axis lies above the profiled sheeting.

9.7.2(5)

Therefore the sagging bending moment resistance should be determined from the stress distribution shown in the figure below.

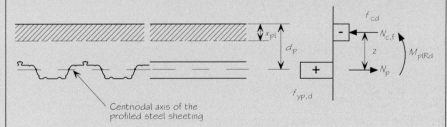

Centriodal axis of the profiled steel sheeting

The depth of concrete in compression is:

$$x_{pl} = \frac{A_{pe} f_{yp.d}}{b f_{cd}}$$

where:

b is the width of the floor slab being considered, here;

$b = 1000$ mm

$$x_{pl} = \frac{1424 \times 350}{1000 \times 14.2} = 35.1 \text{ mm}$$

Bending resistance – full shear connection

For full shear connection, the design moment resistance is:

$$M_{pl.Rd} = A_p f_{yd} \left(d_p - x_{pl}/2 \right)$$

$d_p = h -$ depth from soffit to centroidal axis of sheeting

$d_p = 130 - 30.5 = 99.5$ mm

The plastic bending resistance per metre width of the slab is:

$$M_{pl.Rd} = 1424 \times 350 \times (99.5 - 35.1/2) \times 10^{-6} = 40.84 \text{ kNm/m}$$

$M_{pl.Rd} = 40.84$ kNm/m

$$\frac{M_{Ed}}{M_{pl.Rd}} = \frac{9.35}{40.84} = 0.23 < 1.0$$

Therefore the bending moment resistance for full shear connection is adequate.

| Example 08 Composite slab | Sheet | 6 | of | 7 | Rev |

9.7.3

Longitudinal shear resistance: m-k method

The method given in 9.7.3 may be used to determine the design resistance to longitudinal shear ($V_{l,Rd}$). In this example, the benefits of end anchorage have been ignored.

9.7.3(4)

$$V_{l,Rd} = \frac{bd_p}{\gamma_{vs}}\left(\frac{mA_p}{bL_s} + k\right)$$

m and k are design values obtained from the manufacturer. For the CF60 steel deck the following values have been obtained from the output from the software *Comdek*.[2]

$m = 157.2$ N/mm²

$k = 0.1232$ N/mm²

9.7.3(5)

For a uniform load applied to the whole span length;

$$L_s = \frac{L}{4} = \frac{3000}{4} = 750 \text{ mm}$$

9.7.3(4)

$$V_{l,Rd} = \left[\frac{1000\times99.5}{1.25}\times\left(\frac{157.2\times1424}{1000\times750}+0.1232\right)\right]\times10^{-3}$$

$$= 33.56 \text{ kN/m}$$

$V_{Ed} = 13.07$ kN/m

The design shear resistance must not be less than the maximum design vertical shear.

$$\frac{V_{Ed}}{V_{l,Rd}} = \frac{13.07}{33.56} = 0.39 < 1.0$$

Therefore the design resistance to longitudinal shear is adequate.

Design vertical shear resistance

The vertical shear resistance will normally be based on BS EN 1992-1-1 Equation 6.2b. Using the nomenclature in BS EN 1994-1-1, the equation becomes:

$$V_{v,Rd} = \left(v_{min} + k_1\sigma_{cp}\right)b_s d_p$$

Although in reality the slab is continuous, it is normally convenient to design it as simply supported (except for the fire condition when the benefits of continuity are usually recognised). As a consequence of this, the beneficial effect of the hogging moments at the supports is neglected, such that $\sigma_{cp} = 0$. Hence,

$$V_{v,Rd} = v_{min}b_s d_p$$

2) If the m and k values are not available from the manufacturer, the longitudinal shear for slabs without end anchorage may be determined using the partial connection method given in 9.7.3(8) of BS EN 1994-1-1, which requires the shear bond values

| Example 08 Composite slab | Sheet | 7 | of | 7 | Rev |

<table>
<tr><td>

BS EN 1992-
1-1
NA 2
Table NA.1

</td><td>

The recommended value of v_{min} is

$$v_{min} = 0.035 k^{3/2} f_{ck}^{1/2}$$

where $k = 1 + \sqrt{200/d_p} \leq 2.0$

$1 + \sqrt{200/99.5} = 2.42$, so $k = 2.0$

$v_{min} = 0.035 \times 2^{3/2} \times 25^{1/2} = 0.49 \, N/mm^2$

$V_{v,Rd} = 0.49 \times 99.5 = 48.8 \, kN/m, > 13.19 \, kN/m$, OK

Therefore the vertical shear resistance is satisfactory.

All design checks of the composite slab in the ultimate limit state are satisfied.

Serviceability limit state (SLS):

The serviceability limit state checks for the composite slab are not given in this example. Some notes are given below.

</td></tr>
</table>

9.8.1 (2)

Cracking of concrete

As the slab is designed as being simply supported, only anti-crack reinforcement is needed. The cross-sectional area of the reinforcement (A_s) above the ribs of the profiled steel sheeting should not be less than 0.4% of the cross-sectional area of the concrete above the ribs for unpropped construction. Crack widths may still need to be verified in some circumstances.

Deflection:

9.8.2(5)

For an internal span of a continuous slab the vertical deflection may be determined using the following approximations:

- the second moment of area may be taken as the average of the values for the cracked and un-cracked section;

- for concrete, an average value of the modular ratio, n, for both long-term and short-term effects may be used.

Fire

This example has not considered fire resistance, which will sometimes govern the design

Job No.	Sheet *1* of *11*	Rev C
Job Title Example no. 09	Revised by DGB, April 09	
Subject Bracing and bracing connections		

Silwood Park, Ascot, Berks SL5 7QN
Telephone: (01344) 636525
Fax: (01344) 636570

CALCULATION SHEET

Client		Made by	JPR	Date	Nov 2006
		Checked by	AGK	Date	Dec 2006

PUnless stated otherwise all references are to BS EN 1993-1-1

Bracing and bracing connections

Design summary:

(a) The wind loading at each floor is transferred to two vertically braced end bays on grid lines 'A' and 'J' by the floors acting as diaphragms.

(b) The bracing system must carry the equivalent horizontal forces (EHF) in addition to the wind loads.

(c) Locally, the bracing must carry additional loads due to imperfections at splices (cl 5.3.3(4)) and restraint forces (cl 5.3.2(5)). These imperfections are considered in turn in conjunction with external lateral loads but not at the same time as the EHF.

(d) The braced bays, acting as vertical pin-jointed frames, transfer the horizontal wind load to the ground.

(e) The beams and columns that make up the bracing system have already been designed for gravity loads[1]. Therefore, only the diagonal members have to be designed and only the forces in these members have to be calculated.

(f) All the diagonal members are of the same section, thus, only the most heavily loaded member has to be designed.

Forces in the bracing system

BS EN 1991-1-4

Total overall unfactored wind load[2], F_w = 925 kN

With two braced bays, total unfactored load to be resisted by each braced bay = 0.5 × 925 = 463 kN

Actions

Roof

Permanent action   = 0.9 kN/m²

Variable action   = 0.6 kN/m²

Floor

Permanent action   = 3.7 kN/m²

Variable action   = 3.3 kN/m²

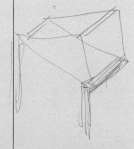

---

1) It should be checked that these members can also carry any loads imposed by the wind when they form part of the bracing system, considering the appropriate combination of actions.

2) In this example, the wind load considered is only for the direction shown on structural arrangement and loading, Sheet 2. In practice, other directions must also be considered.

| Example 09 Bracing and bracing connections | Sheet 2 of 11 | Rev |

Ultimate Limit State (ULS)

BS EN 1990
NA 2.2.3.2
Table
NA.A1.2(B)

Partial factors for actions

Partial factor for permanent actions $\quad \gamma_G = 1.35$

Partial factor for variable actions $\quad \gamma_Q = 1.5$

Reduction factor $\quad \xi = 0.925$

BS EN 1990
NA 2.2.2
Table NA.A1.1

ψ_0 factors

For imposed floor loads (office areas) $\quad \psi_0 = 0.7$

For snow loads on roofs (H ≤ 1000m a.s.l) $\psi_0 = 0.5$

Combinations of actions for ULS, using Eqn 6.10b

Design value of combined actions

$$= \xi\gamma_G G_k + \gamma_Q Q_k + \psi_0 \gamma_Q Q_k$$

In this example, the bracing will be verified for one design case, using Equation 6.10b, with wind as the leading variable action. The Equivalent horizontal forces (EHF) will also be calculated for this combination. In practice, Equation 6.10a should also be checked, and additional combinations (for example with the imposed floor load as the leading variable action).

Design wind load at ULS

Using Equation 6.10b with wind as the leading variable action, the design wind load per braced bay is:

$$F_{Ed} = 1.5 \times 463 = 695 \text{ kN}$$

Distributing this total horizontal load as point loads at roof and floor levels, in proportion to the storey heights:

Roof level $\qquad \dfrac{2.25}{18.5} \times 695 = 85$ kN

3rd & 2nd floor levels $\qquad \dfrac{4.5}{18.5} \times 695 = 169$ kN

1st floor level $\qquad \dfrac{4.75}{18.5} \times 695 = 178$ kN

Ground at column base level $\qquad \dfrac{2.5}{18.5} \times 695 = 94$ kN*

*Assume that this load is taken out in shear through the ground slab and is therefore not carried by the frame.

| Example 09 Bracing and bracing connections | Sheet *3* of *11* | Rev |

Equivalent horizontal forces

With wind as the leading variable action, the design values of the combined floor and roof actions are:

Design value for combined roof actions

$$= 0.925 \times 1.35 \times 0.9 + 1.5 \times 0.5 \times 0.6 = 1.57 \text{ kN/m}^2$$

Design value for combined floor actions

$$= 0.925 \times 1.35 \times 3.7 + 1.5 \times 0.7 \times 3.3 = 8.09 \text{ kN/m}^2$$

Total roof load $= 1.57 \times 14 \times 48 = 1055 \text{ kN}$
Total floor load $= 8.09 \times 14 \times 48 = 5437 \text{ kN}$

Equivalent horizontal forces for each bracing system are:

roof level $= \dfrac{1055}{200} \times 0.5 = 2.64 \text{ kN}$

floor level $= \dfrac{5437}{200} \times 0.5 = 13.6 \text{ kN}$

BS EN 1993-1-1 5.3.2(3)

The equivalent horizontal forces have been based on $\phi_0 = 1/200$. The α_h and α_m factors have not been used, which is conservative. Use of the α_h and α_m factors will reduce the forces in the bracing.

Horizontal forces at ground level

Horizontal design force due to wind
$$= (85 + 169 + 169 + 178) = 601 \text{ kN}$$

Horizontal design force due to equivalent horizontal loads
$$= 2.64 + 3 \times 13.6 = 43.4 \text{ kN}$$

Total horizontal design force per bracing system
$$= 601 + 43.4 = 644.4 \text{ kN}$$

A computer analysis of the bracing system can be performed to obtain the member forces. Alternatively, hand calculations can be carried out to find the member forces. Simply resolving forces horizontally at ground level is sufficient to calculate the force in the lowest (most highly loaded) bracing member, as shown in Figure 9.1.

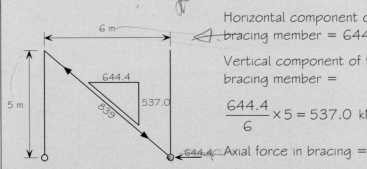

Horizontal component of force in bracing member = 644.4 kN

Vertical component of force in bracing member =

$$\frac{644.4}{6} \times 5 = 537.0 \text{ kN}$$

Axial force in bracing =

$$\sqrt{644.4^2 + 537.0^2} = 839 \text{ kN}$$

Figure 9.1 Lowest bracing

| Example 09 Bracing and bracing connections | Sheet | 4 | of | 11 | Rev |

	Partial factors for resistance
NA 2.15	γ_{M0} = 1.0
BS EN 1993-1-8 NA 2.3 Table NA.1	γ_{M1} = 1.0
	γ_{M2} = 1.25 (for bolts and welds)

Trial section

Try: 219.1 × 10.0 mm thick Circular Hollow Section (CHS), grade S355

Section Properties

SCI P363

Area	A	= 65.7 cm²
Second moment of area	I	= 3600 cm⁴
Radius of gyration	i	= 7.40 cm
Thickness	t	= 10.0 mm
Ratio for local Buckling	d/t	= 21.9

Material properties

NA 2.4
BS EN 10210-1
Table A3
3.2.6 (1)

As $t \leq 16$ mm, for S355 steel

Yield strength	f_y	= 355 N/mm²
modulus of elasticity	E	= 210 kN/mm²

5.5
Table 5.2

Section classification

Class 1 limit for section in compression, $d/t \leq 50 \varepsilon^2$

$\varepsilon = (235/f_y)^{0.5}$, f_y = 355 N/mm², ε = 0.82

$d/t \leq 50e^2 = 50 \times 0.82^2 = 33.6$

Since 21.9 < 33.6, the section is Class 1 for axial compression

Design of member in compression

Cross sectional resistance to axial compression

6.2.4(1)
Eq. 6.9

Basic requirement $\quad \dfrac{N_{Ed}}{N_{c,Rd}} \leq 1.0$

N_{Ed} is the design value of the applied axial force

N_{Ed} = 839 kN

$N_{c,Rd}$ is the design resistance of the cross-section for uniform compression

6.2.4(2)
Eq. 6.10

$N_{c,Rd} = \dfrac{A \times f_y}{\gamma_{M0}}$ (For Class 1, 2 and 3 cross-sections)

$N_{c,Rd} = \dfrac{6570 \times 355}{1.0} \times 10^{-3} = 2332$ kN

| Example 09 Bracing and bracing connections | Sheet | 5 | of | 11 | Rev |

$$\frac{N_{Ed}}{N_{c,Rd}} = \frac{839}{2332} = 0.36 \ < \ 1.0$$

Therefore, the resistance of the cross section is adequate.

Flexural buckling resistance

6.3.1.1(1)
Eq. 6.46

For a uniform member under axial compression the basic requirement is:

$$\frac{N_{Ed}}{N_{b,Rd}} \leq 1.0$$

$N_{b,Rd}$ is the design buckling resistance and is determined from:

6.3.1.1(3)
Eq. 6.47

$$N_{b,Rd} = \frac{\chi A f_y}{\gamma_{M1}} \text{ (For Class 1, 2 and cross-sections)}$$

6.3.1.2(1)

χ is the reduction factor for buckling and may be determined from Figure 6.4.

Table 6.2

For hot finished CHS in grade S355 steel use buckling curve '*a*'

Use buckling curve '*a*'

For flexural buckling the slenderness is determined from:

6.3.1.3(1)
Eq. 6.50

$$\overline{\lambda} = \sqrt{\frac{A f_y}{N_{cr}}} = \left(\frac{L_{cr}}{i}\right)\left(\frac{1}{\lambda_1}\right) \text{ (For Class 1, 2 and 3 cross-sections)}$$

where:

L_{cr} is the buckling length

As the bracing member is pinned at both ends, conservatively take:

$$L_{cr} = L = \sqrt{5000^2 + 6000^2} = 7810 \text{ mm}$$

$L_{cr} = 7810$ mm

$$\lambda_1 = 93.9\varepsilon$$

Table 5.2

$$\varepsilon = \sqrt{\frac{235}{f_y}} = \sqrt{\frac{235}{355}} = 0.81$$

$$\lambda_1 = 93.9 \times 0.81 = 76.1$$

$\lambda_1 = 76.1$

6.3.1.3(1)
Eq. 6.50

$$\overline{\lambda} = \left(\frac{7810}{74}\right) \times \left(\frac{1}{76.1}\right) = 1.39$$

$\overline{\lambda} = 1.39$

Figure 6.4

For $\overline{\lambda} = 1.39$ and buckling curve '*a*'

$$\chi = 0.42$$

Therefore,

$\chi = 0.42$

6.3.1.1(3)
Eq. 6.47

$$N_{b,Rd} = \frac{0.42 \times 65.7 \times 10^2 \times 355}{1.0} \times 10^{-3} = 980 \text{ kN}$$

Flexural buckling resistance
$N_{b,Rd} = 980$ kN

6.3.1.1(1)
Eq. 6.46

$$\frac{N_{Ed}}{N_{b,Rd}} = \frac{839}{980} = 0.86 \ < \ 1.0$$

Therefore, the flexural buckling resistance of the section is adequate.

| Example 09 Bracing and bracing connections | Sheet | 6 | of | 11 | Rev |

6.2.3	### Design of member in tension

When the wind is applied in the opposite direction, the bracing member considered above will be loaded in tension. By inspection, the tensile capacity is equal to the cross-sectional resistance, 2332 kN, > 839 kN, OK

Resistance of connection (see Figure 9.2)

Assume the CHS is connected to the frame via gusset plates. Flat end plates fit into slots in the CHS section and are fillet welded to the CHS. Bolts in clearance holes transfer the load between the end plate and gusset plates.

Verify the connection resistance for 839 kN tensile force.

Try: 8 No non-preloaded Class 8.8 M24 diameter bolts in 26 mm diameter clearance holes

P363

Assume shear plane passes through the threaded part of the bolt

Cross section area, $\quad\quad\quad A \quad = A_s = 353$ mm^2
Clearance hole diameter, $\quad\quad d_o \quad = 26$ mm

BS EN 1993-1-8
Table 3.1

For Class 8.8 non-preloaded bolts:

Yield strength $\quad\quad\quad\quad\quad\quad f_{yb} \quad = 640$ N/mm^2
Ultimate tensile strength $\quad\quad\quad f_{ub} \quad = 800$ N/mm^2

Positioning of holes for bolts:

(Minimum) End distance (e_1) $\quad$ 1.2 d_o = 31.2 mm < e_1 $\quad$ = 40 mm
(Minimum) Edge distance (e_2) $\quad$ 1.2 d_o = 31.2 mm < e_2 $\quad$ = 60 mm
(Minimum) Spacing (p_1) $\quad\quad$ 2.2 d_o = 57.2 mm < p_1 $\quad$ = 80 mm
(Minimum) Spacing (p_2) $\quad\quad$ 2.4 d_o = 62.4 mm < p_2 $\quad$ = 130 mm

(Maximum) e_1 & e_2,
larger of $\quad$ 8t = 120 mm or 125 mm > 40 mm & 60 mm

(Maximum) p_1 & p_2
Smaller of $\quad$ 14t = 210 mm or 200 mm > 80 mm & 130 mm
Therefore, bolt spacings comply with the limits.

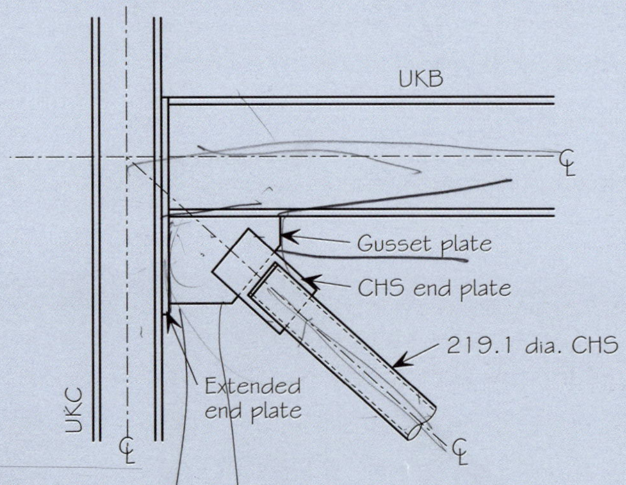

Figure 9.2 Bracing setting out and connection detail

| Example 09 Bracing and bracing connections | Sheet | 7 | of | 11 | Rev |

BS EN
1993-1-8
Figure 3.1

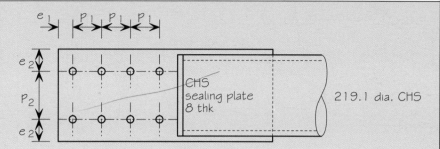

Positioning of holes for bolts

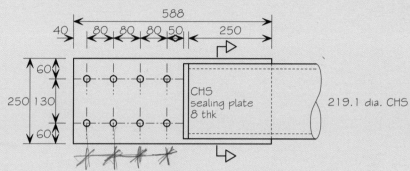

Grade S275 end plate 588 × 250 × 15 mm thick to fit into a slotted hole in the CHS

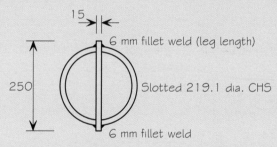

Figure 9.3 CHS end plate details

Shear resistance of bolts

BS EN
1993-1-8
Table 3.4

The resistance of a single bolt in shear is determined from:

$$F_{v,Rd} = \frac{\alpha_v f_{ub} A}{\gamma_{M2}} = \frac{0.6 \times 800 \times 353}{1.25} \times 10^{-3} = 135 \text{ kN}$$

Where:

$\alpha_v = 0.6$ for grade 8.8 bolts

Minimum No of bolts required is $\dfrac{N_{Ed}}{F_{v,Rd}} = \dfrac{839}{135} = 6.2$ bolts

Therefore, provide 8 bolts in single shear.

Bearing resistance of bolts

Assume gusset plate has a thickness no less than the 15 mm end plate.

Resistance of a single bolt to shear: $F_{v,Rd} = 135$ kN

77

| Example 09 Bracing and bracing connections | Sheet 8 of 11 | Rev |

BS EN 1993-1-1 NA 2.4 BS EN 10025-2 Table 7	End plate is a grade S275 and as $t \leq 16$ mm, for S275 steel Yield strength $\quad\quad f_y = 275$ N/mm² as $3 \leq t \leq 100$ mm; Ultimate tensile strength $\quad f_u = 410$ N/mm²	
BS EN 1993-1-8 Table 3.4	The bearing resistance of a single bolt is determined from: $$F_{b,Rd} = \frac{k_1 \alpha_b f_u dt}{\gamma_{M2}}$$ α_b is the least value of α_d, $\dfrac{f_{ub}}{f_{u,p}}$ and 1.0 For end bolts $\alpha_d = \dfrac{e_1}{3d_o} = \dfrac{40}{3 \times 26} = 0.51$ For inner bolts $\alpha_d = \dfrac{p_1}{3d_o} - \dfrac{1}{4} = \left(\dfrac{80}{3 \times 26}\right) - \left(\dfrac{1}{4}\right) = 0.78$ $\dfrac{f_{ub}}{f_{u,p}} = \dfrac{800}{410} = 1.95$ Therefore: For end bolts $\quad\quad\quad\quad \alpha_b = 0.51$ For inner bolts $\quad\quad\quad\quad \alpha_b = 0.78$ Conservatively consider $\quad \alpha_b = 0.51$ for each bolt. For edge bolts k_1 is the smaller of $2.8\dfrac{e_2}{d_o} - 1.7$ or 2.5 $\left(2.8 \times \dfrac{60}{26}\right) - 1.7 = 4.8$ For inner bolts k_1 is the smaller of $1.4\dfrac{p_2}{d_o} - 1.7$ or 2.5 $\left(1.4 \times \dfrac{130}{26}\right) - 1.7 = 5.3$ Therefore: For both end and inner bolts $\quad\quad k_1 = 2.5$ The least bearing resistance of a single bolt in this connection is thus: $$F_{b,Rd} = \frac{2.5 \times 0.51 \times 410 \times 24 \times 15}{1.25} \times 10^{-3} = 151 \text{ kN}$$ Resistance of all six bolts in bearing may be conservatively taken as: $8F_{b,Rd} = 8 \times 151 = 1208$ kN	$\alpha_b = 0.51$ $k_1 = 2.5$ $F_{b,Rd} = 151$ kN Resistance of 8 bolts in bearing 1208 kN

| Example 09 Bracing and bracing connections | Sheet | 9 | of | 11 | Rev |

BS EN 1993-1-8 3.7	**Group of fasteners** Because the shear resistance of the bolts (135 kN) is less than the minimum bearing resistance of any bolt (151 kN), the design resistance of the group is taken as: $8 \times 135 = 1080$ kN	Resistance of the bolt group 1080 kN
	Design of fillet weld (see Figure 9.3)	
BS EN 1993-1-8 4.5	Assume 6 mm leg length fillet weld is used on both sides, top and bottom, of the fitted end plate. Use the simplified method in 4.5.3.3	
BS EN 1993-1-8 4.5.3.3(3)	Design shear strength, $f_{vw,d} = \dfrac{f_u / \sqrt{3}}{\beta_w \gamma_{M2}}$	
BS EN 1993-1-8 Table 4.1	Correlation factor, for S275 steel $\beta_w = 0.85$ Throat thickness of weld $a = 0.7 \times \text{leg length} = 0.7 \times 6.0 = 4.2$ mm Therefore, $f_{vw,d} = \dfrac{410 / \sqrt{3}}{0.85 \times 1.25} = 222.8$ N/mm^2	
BS EN 1993-1-8 4.5.3.3(2) See sheet 7	Design resistance of weld per unit length is: $F_{vw,d} = f_{vw,d} a = 222.8 \times 4.2 = 935.8$ N/mm	
	Hence, for four welds, each with an effective length of: $l_{eff} = 250 - (2 \times 6.0) = 238$ mm the shear resistance is $4 F_{w,Rd} l_{eff} = 4 \times 935.8 \times 238 \times 10^{-3} = 891$ kN, > 839 kN, OK	Shear resistance of 4 by 238 mm long 6 mm fillet welds is: 891 kN
	Local resistance of CHS wall In the absence of guidance in BS EN 1993-1-1 for the shear area of a plain rectangular area, it is assumed that the shear area, $A_v = 0.9dt$, where d is the depth of the rectangular area and t the thickness. Total shear area $= 4 \times 0.9 \times 250 \times 10 = 9000$ mm^2 Shear resistance is $\dfrac{A\left(f_y / \sqrt{3}\right)}{\gamma_{M0}} = \dfrac{900 \times 355 / \sqrt{3}}{1.0 \times 10^3} = 1844$ kN 1844 kN > 839 kN, OK	 

| Example 09 Bracing and bracing connections | Sheet | *10* of *11* | Rev |

Tensile resistance of end plate (see Figure 9.4)

Two modes of failure are to be considered:
i) cross-sectional failure and
ii) block tearing failure.

BS EN
1993-1-8
3.10.2

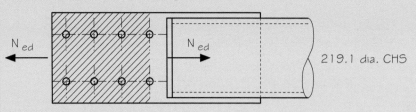

i) Cross-sectional failure

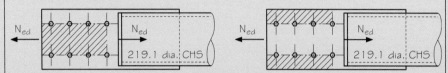

ii) Block tearing failure

Figure 9.4 Plate failure modes

i) Cross-sectional failure

6.2.3(1)

Basic requirement: $\dfrac{N_{Ed}}{N_{t,Rd}} \leq 1.0$

6.2.3(2)

For a cross-section with holes, the design tensile resistance is taken as the smaller of $N_{pl,Rd}$ and $N_{u,Rd}$:

Eqn. 6.6

$N_{pl,Rd} = \dfrac{A \times f_y}{\gamma_{M0}}$

A is the gross cross-sectional area:

$A = 250 \times 15 = 3750$ mm^2

$N_{pl,Rd} = \dfrac{3750 \times 275}{1.0} \times 10^{-3} = 1031$ kN, > 839 kN, OK

$N_{pl,Rd} = 1031$ kN

Eqn. 6.7

$N_{u,Rd} = \dfrac{0.9 \times A_{net} \times f_u}{\gamma_{M2}}$

6.2.2.2

$A_{net} = 3750 - (2 \times 26 \times 15) = 2970$ mm^2

$N_{u,Rd} = \dfrac{0.9 \times 2970 \times 410}{1.25} \times 10^{-3} = 876$ kN, > 839 kN, OK

$N_{u,Rd} = 876$ kN

ii) Block tearing failure

BS EN
1993-1-8
3.10.2 (2)

For a symmetric bolt group subject to concentric loading, the design block tearing resistance ($V_{Eff,1,Rd}$) is determined from:

$V_{eff,1,Rd} = \dfrac{f_u A_{nt}}{\gamma_{M2}} + (1/\sqrt{3})\dfrac{f_y A_{nv}}{\gamma_{M0}}$

where:

A_{nt} is the net area subject to tension

A_{nv} is the net area subject to shear

| Example 09 Bracing and bracing connections | Sheet *11* of *11* | Rev |

A_{nt} is minimum of $(p_2 - d_o) t_p$ and $2 (e_z - 0.5 d_o) t_p$

$(p_2 - d_o) t_p = (130 - 26) \times 15 = 1560 \text{ mm}^2$

$2 (e_2 - 0.5 d_o) t_p = 2 (60 - 13) \times 15 = 1410 \text{ mm}^2$

$A_{nt} = 1410 \text{ mm}^2$

$A_{nv} = 2(3 p_1 + e_1 \quad 2.5 d_0) t_w = 2 \times 215 \times 15 = 6450 \text{ mm}^2$

$$V_{eff,1,Rd} = \frac{410 \times 1410}{1.25 \times 10^3} + \left(1/\sqrt{3}\right) \times \frac{275 \times 6450}{1.0 \times 10^3} = 1487 \text{ kN}$$

$V_{Eff,1,Rd} = 1487 \text{ kN}$

1487 kN > 839 kN, OK

The gusset plates would also require checking for shear, bearing and welds, together with full design check for the extended beam end plates[3].

---

3) The gusset plate must be checked for yielding across an effective dispersion width of the plate. When the bracing member is in compression, buckling of the gusset plate must be prevented and therefore a full design check must be carried out.

Job No.		Sheet	*1*	of *7*	Rev	C
Job Title	Example no. 10		Revised by DGB, April 09			
Subject	Beam-to-column flexible end plate connection					
Client		Made by	MS	Date	Nov 2006	
		Checked by	PA	Date	Dec 2006	

Unless stated otherwise all references are to BS EN 1993-1-8:2005

BS EN 1993-1-1 NA 2.4

BS EN 10025-2 Table 7

BS EN 1993-1-1 NA 2.15

Access steel document SN013a-EN-EU

Beam-to-column flexible end plate connection

Design the beam-to-column connection at level 1 between gridlines G and 2.

Initial sizing of the components of the connection

Column   254 × 254 × 73 UKC in S275 steel

Beam   457 × 191 × 82 UKB in S275 steel

For the beam, $f_y = 275$ N/mm²; $f_u = 410$ N/mm²; $h_b = 460$mm;
$t_w = 9.9$mm; $t_f = 16$mm

For the plate, $f_u = 410$ N/mm²

$$V_{cRd} \approx \frac{h_b \times t_w \times \left(\frac{f_y}{\sqrt{3}}\right)}{\gamma_{M0}} = \frac{460 \times 9.9 \times \left(\frac{275}{\sqrt{3}}\right)}{1.0 \times 10^{-3}} = 723\text{kN}$$

From example 1 the design shear force at ULS, $V_{Ed} = 230$ kN

$V_{Ed} = 230$ kN

Because $230 < 0.75\,V_{cRd}$, a partial depth endplate is proposed.

$h_b < 500$ mm, so 8 or 10 mm endplate is proposed.

End plate depth is minimum $0.6\,h_b = 276$ mm; propose 280 mm.

Assuming M20 bolts, number of bolts = 239/74 = 3.2

6 M20 bolts are proposed.

Based on the above, the initial sizing of the connection components is shown in Figure 10.1.

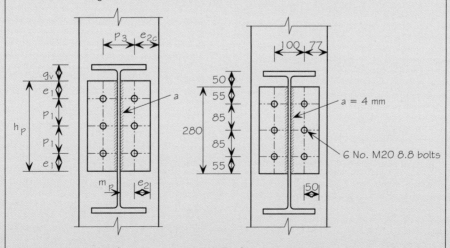

(a) parameters definition   (b) Profile adopted based on initial sizing

Figure 10.1   Connection details

Bolt details

The bolts are fully threaded, non-preloaded, M20 8.8, 60 mm long, as generally used in the UK.

Tensile stress area of bolt $\quad A_s \quad = 245$ mm^2

Diameter of the holes $\quad\quad d_0 \quad = 22$ mm

Diameter of the washer $\quad\quad d_w \quad = 37$ mm

Yield strength $\quad\quad\quad\quad f_{yb} \quad = 640$ N/mm^2

Ultimate tensile strength $\quad f_{ub} \quad = 800$ N/mm^2

3.5, Table 3.3 · Limits for locations and spacings of bolts

End distance $\quad e_1 \quad = 55$ mm

Minimum $= 1.2d_0 = 1.2 \times 22 \quad = 26.4$ mm $\quad < 55$ mm, OK

Edge distance $\quad e_2 \quad = 50$ mm

Limits are the same as those for end distance.

Minimum $= 1.2d_0 = 1.2 \times 22 \quad = 26.4$ mm $\quad < 50$ mm, OK

Spacing (vertical pitch) $\quad\quad p_1 \quad = 85$ mm

Minimum $= 2.2d_0$

$2.2d_0 = 2.2 \times 22 = 48.4$ mm < 85 mm, OK

$14t_p = 14 \times 10 = 140$ mm > 85 mm

Spacing (horizontal gauge) $\quad p_3 \quad = 100$ mm

Minimum $= 2.4d_0$

$2.4d_0 = 2.4 \times 22 = 52.8$ mm < 100 mm, OK

4.7.3

Weld design

Access Steel document SN014a-EN-EU

For full strength "side" welds

Throat (a) $\geq 0.39 \times t_w$

$a \geq 0.39 \times 9.9 = 3.86$ mm; adopt throat (a) of 4mm, leg $= 6$ mm

Weld throat thickness, $a = 4$ mm Leg $= 6$ mm

BS EN 1993-1-1 6.1(1) Table 2.1 BS EN 1993-1-8 NA 2.3 Table NA.1

Partial factors for resistance

$\gamma_{M0} \quad = 1.0$

$\gamma_{M2} \quad = 1.25$ (for shear)

$\gamma_{M2} \quad = 1.1$ (for bolts in tension)

Access Steel document SN018a-EN-EU

$\gamma_{Mu} \quad = 1.1$

The partial factor for resistance γ_{Mu} is used for the tying resistance. Elastic checks are not appropriate; irreversible deformation is expected.

SN014a- EN-EU	The connection detail must be ductile to meet the design requirement that it behaves as nominally pinned. For the UK, and based on SN014, the ductility requirement is satisfied if the supporting element (column flange in this case) or the end plate, complies with the following conditions: $$t_p \leq \frac{d}{2.8}\sqrt{\frac{f_{u,b}}{f_{y,p}}} \text{ or } t_{f,c} \leq \frac{d}{2.8}\sqrt{\frac{f_{u,b}}{f_{y,c}}}$$ $$\frac{d}{2.8}\sqrt{\frac{f_{u,b}}{f_{y,p}}} = \left(\frac{20}{2.8}\right) \times \sqrt{\frac{800}{275}} = 12 \text{ mm}$$ Since $t_p = 10$ mm < 12 mm, ductility is ensured. ### Joint shear resistance The following table gives the complete list of design resistances that need to be determined for the joint shear resistance. Only the critical checks are shown in this example. The critical checks are denoted with an * in the table. Because a full strength weld has been provided, no calculations for the weld are required.	

Mode of failure	
Bolts in shear*	$V_{Rd,1}$
End plate in bearing*	$V_{Rd,2}$
Supporting member (column) in bearing	$V_{Rd,3}$
End plate in shear (gross section)	$V_{Rd,4}$
End plate in shear (net section)	$V_{Rd,5}$
End plate in block shear	$V_{Rd,6}$
End plate in bending	$V_{Rd,7}$
Beam web in shear*	$V_{Rd,8}$

3.6.1 & Table 3.4	### Bolts in shear Assuming the shear plane passes through the threaded portion of the bolt, the shear resistance $F_{v,Rd}$ of a single bolt is given by: $$F_{v,Rd} = \frac{\alpha_v f_{ub} A}{\gamma_{M2}}$$	
Access Steel document SN014a-EN- EU	Although not required by the Eurocode, a factor of 0.8 is introduced into the above equation, to allow for the presence of modest tension (not calculated) in the bolts. For bolt class 8.8, $\alpha_v = 0.6$, therefore, $$F_{v,Rd} = 0.8 \times \frac{0.6 \times 800 \times 245 \times 10^{-3}}{1.25} = 75.2 \text{ kN}$$ For 6 bolts, $V_{Rd,1} = 6 \times 75.2 = 451$ kN	$V_{Rd,1} = 451$ kN

3.6.1 & Table 3.4

End plate in bearing

The bearing resistance of a single bolt, $F_{b,Rd}$ is given by:

$$F_{b,Rd} = \frac{k_1 \alpha_b f_{u,p} d t_p}{\gamma_{M2}}$$

Where:

$$\alpha_b = \min\left(\alpha_d ; \frac{f_{u,b}}{f_{u,p}} ; 1.0\right)$$

and $a_d = \frac{e_1}{3d_o}$ for end bolts and $\frac{p_1}{3d_o} - \frac{1}{4}$ for inner bolts

For end bolts,

$$\alpha_b = \min\left(\frac{55}{3\times22} ; \frac{800}{410} ; 1.0\right) = \min(0.83 ; 1.95 ; 1.0) = 0.83$$

For inner bolts,

$$\alpha_b = \min\left(\frac{85}{3\times22} - \frac{1}{4} ; \frac{800}{410} ; 1.0\right) = \min(1.04 ; 1.95 ; 1.0) = 1.0$$

$$k_1 = \min\left(2.8\frac{e_2}{d_o} - 1.7 ; 2.5\right) = \min\left(2.8\times\left(\frac{50}{22}\right) - 1.7 ; 2.5\right)$$

Therefore, k_1 = minimum (4.66; 2.5) = 2.5

Therefore, for the end bolts,

$$F_{b.Rd} = \frac{2.5\times0.83\times410\times20\times10}{1.25}\times10^{-3} = 136.1\text{kN}$$

And for the inner bolts,

$$F_{b.Rd} = \frac{2.5\times1.0\times410\times20\times10}{1.25}\times10^{-3} = 164.0\text{kN}$$

The bearing resistance of the bolts

= 2 × 136.1 + 4 × 164 = 928 kN

$V_{Rd,2}$ = 928 kN

BS EN 1993-1-8:2005 3.7

Group of Fasteners

Because the shear resistance of the fasteners (75.2 kN) is less than the bearing resistance, the resistance of the group of fasteners must be taken as the number of fasteners multiplied by the smallest design resistance of the individual fasteners – in this case 75.2 kN

Resistance of the group = 6 × 75.2 = 451 kN

Beam web in shear

Shear resistance is checked only for the area of the beam web connected to the end plate.

BS EN 1993-1-1 6.2.6(2)

The design plastic shear resistance is given by:

$$V_{pl,Rd} = V_{Rd,8} = \frac{A_v(f_{y,b}/\sqrt{3})}{\gamma_{M0}}$$

Access Steel document SN014a-EN-EU	A factor of 0.9 is introduced into the above equation when calculating the plastic shear resistance of a plate (which is not covered in BS EN 1993-1-1)

$$= 0.9 \times \frac{(280 \times 9.9) \times (275 / \sqrt{3})}{1.0} \times 10^{-3} = 396 \text{ kN}$$

$V_{Rd,8} = 396$ kN

The design shear resistance of the connection is 396 kN, > 230 kN, OK

Tying resistance of end plate [1]

The following table gives the complete list of design resistances that need to be determined for the tying resistance of the end plate. Only critical checks are shown. The critical checks are denoted with an * in the table. The check for bolts in tension is also carried out because the tension capacity of the bolt group is also needed for the end plate in bending check.

Mode of failure	
Bolts in tension	$N_{Rd,u,1}$
End plate in bending*	$N_{Rd,u,2}$
Supporting member in bending	$N_{Rd,u,3}$
Beam web in tension	$N_{Rd,u,4}$

Bolts in tension

3.6.1 & Table 3.4

The tension resistance for a single bolt is given by:

$$F_{t,Rd} = \frac{k_2 f_{ub} A_s}{\gamma_{Mu}}$$

$$k_2 = 0.9$$

$$N_{Rd,u,1} = F_{t,Rd} = \frac{0.9 \times 800 \times 245}{1.1} \times 10^{-3} = 160.4 \text{ kN}$$

For 6 bolts, $N_{Rd,u,1} = 6 \times 160.4 = 962$ kN

$N_{Rd,u,1} = 962$ kN

End plate in bending

Equivalent tee-stub considered for the end plate in bending checks:

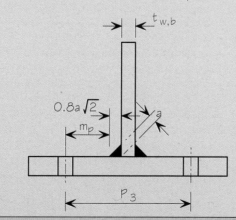

1) The tying force to be resisted should be determined following the guidance in BS EN 1991-1-7 or the applicable National Regulations i.e. Building Regulations.

6.2.4 &
Table 6.2

$N_{Rd,u,2} = \min(F_{Rd,u,ep1}; F_{Rd,u,ep2})$

For mode 1: $F_{Rd,u,ep1} = F_{T,1,Rd} = \dfrac{(8n_p - 2e_w)M_{pl,1,Rd}}{2m_pn_p - e_w(m_p + n_p)}$

For mode 2: $F_{Rd,u,ep1} = F_{T,2,Rd} = \dfrac{2M_{pl,2,Rd} + n_p\sum F_{t,Rd}}{m_p + n_p}$

Where:

$n_p = \min(e_2; e_{2,c}; 1.25m_p)$

$m_p = \dfrac{\left(p_3 - t_{w,b} - 2 \times 0.8a\sqrt{2}\right)}{2}$

$e_w = \dfrac{d_w}{4} = \dfrac{37}{4} = 9.25$ mm

(d_w is the diameter of washer or width across points of bolt head or nut)

Here,

$m_p = \dfrac{\left[100 - 9.9 - \left(2 \times 0.8 \times 4 \times \sqrt{2}\right)\right]}{2} = 40.5$ mm

$1.25m_p = 1.25 \times 40.5 = 50.7$ mm

$n_p = \min(50; 77; 50.7) = 50$ mm

$M_{pl,1,Rd} = \dfrac{1}{4}\dfrac{\sum \ell_{eff,1}\,t_p^2 f_{y,p}}{\gamma_{M0}}$ (Mode 1)

$M_{pl,2,Rd} = \dfrac{1}{4}\dfrac{\sum \ell_{eff,2}\,t_p^2 f_{y,p}}{\gamma_{M0}}$ (Mode 2)

Calculate the effective length of the end plate for mode 1 $\left(\sum \ell_{eff,1}\right)$ and mode 2 $\left(\sum \ell_{eff,2}\right)$.

6.2.6.5 &
Table 6.6

For simplicity, the effective length of the equivalent tee stub, ℓ_{eff} is taken as the length of the plate, i.e. 280 mm

Therefore, $\sum \ell_{eff,1} = \sum \ell_{eff,2} = h_p = 280$ mm

$M_{pl,1,Rd,u} = \dfrac{1}{4}\dfrac{h_p t_p^2 f_{u,p}}{\gamma_{Mu}}$

$M_{pl,1,Rd,u} = \dfrac{1}{4} \times \left(\dfrac{280 \times 10^2 \times 410}{1.1}\right) \times 10^{-6} = 2.61$ kNm

Mode 1:

$F_{T,1,Rd} = \dfrac{\left[(8 \times 50) - (2 \times 9.25)\right] \times 2.61 \times 10^3}{\left[2 \times 40.5 \times 50\right] - \left[9.25 \times (40.5 + 50)\right]} = 310$ kN

Mode 2:

In this case $M_{pl,2,Rd,u} = M_{pl,1,Rd,u}$

$$F_{T,1,Rd} = \frac{\left(2 \times 2.61 \times 10^3\right) + \left(50 \times 962\right)}{40.5 + 50} = 589 \text{ kN}$$

$$F_{T,1,Rd} = \min(310; 589) = 310 \text{ kN}$$

Therefore, $N_{Rd,v,2} = 325$ kN

$N_{Rd,u,2} = 325$ kN

Summary of the results

The following tables give the complete list of design resistances that need to be determined for the tying resistance of the end plate. Only critical checks are shown in this example The critical checks are denoted with an * in the tables.

Joint shear resistance

Mode of failure		Resistance
Bolts in shear*	$V_{Rd,1}$	451 kN
End plate in bearing*	$V_{Rd,2}$	928 kN
Supporting member (column) in bearing	$V_{Rd,3}$	
End plate in shear (gross section)	$V_{Rd,4}$	
End plate in shear (net section)	$V_{Rd,5}$	
End plate in block shear	$V_{Rd,6}$	
End plate in bending	$V_{Rd,7}$	
Beam web in shear*	$V_{Rd,8}$	396 kN

The governing value is the minimum value and therefore

$V_{Rd} = 396$ kN

Tying resistance of end plate

Mode of failure		Resistance
Bolts in tension	$N_{Rd,u,,1}$	962 kN
End plate in bending*	$N_{Rd,u,2}$	310 kN
Supporting member in bending	$N_{Rd,u,3}$	N/A
Beam web in tension	$N_{Rd,u,4}$	

The governing value is the minimum value and therefore

$N_{Rd,u} = 310$ kN

Note that if the column flange is thinner than the end plate, this should be checked for bending.

The tying force has not been calculated, but in some cases would be the same magnitude as the shear force. If the resistance is insufficient, a thicker plate could be used (maximum 12 mm to ensure ductility in this instance), or a full depth end plate, or an alternative connection type such as a fin plate.

Unless stated otherwise all references are to BS EN 1993-1-1 and its National Annex	<u>Column base connection</u>

Design conditions for column G2

From previous calculations the following design forces should be considered.

The column is assumed to be pin-ended. However it is crucial the column is stable during the erection phase therefore 4 bolts outside the column profile should be used.

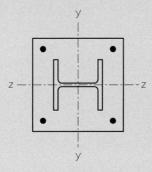

Figure 11.1   Plan of baseplate

Example 05

Characteristic force due to permanent action, $F_{G,k}$ = 466 kN

Characteristic force due to variable action, $F_{Q,k}$ = 416 kN

<u>Ultimate Limit State (ULS)</u>

Partial factors for actions

BS EN 1990-1-1 NA 2.2.3.2 Table NA A1.2(B)

Partial factor for permanent action γ_G = 1.35

Partial factor for variable action γ_Q = 1.5

Reduction factor ξ = 0.925

Combination of actions for ULS

Design value of combined actions

N_{Ed} = 0.925 × 1.35 × 466 + 1.5 × 416 = 1206 kN

Axial force N_{Ed} = 1206 kN

Column details

P 363

Column G2 is a typical internal column

Serial size   254 × 254 × 73 UKC in S275 steel

Height of section	h	= 254.1 mm
Breadth of section	b	= 254.6 mm
Thickness of flange	t_f	= 14.2 mm
Thickness of web	t_w	= 8.6 mm
Cross sectional Area	A	= 93.1 cm²
Section perimeter		= 1490 mm

| Example 11 Column base | | Sheet 2 | of 2 | Rev |

Partial factors for resistance

NA 2.15

$\gamma_{MO} = 1.0 \qquad \gamma_{M2} = 1.25$

Base plate details

BS EN
1992-1-1
Table 3.1

Strength of foundation concrete to be C25/30
(i.e. $f_{ck} = 30$ N/mm²)

BS EN
1992-1-1
NA 2
Table NA 1

$f_{cd} = \dfrac{\alpha_{cc} f_{ck}}{\gamma_c}$

α_{cc} to be taken as 0.85 for axial loading

BS EN
1992-1-1
NA 2
Table.NA 1

$\gamma_C = 1.5$

$f_{cd} = \dfrac{0.85 \times 30}{1.5} = 17$ N/mm²

$f_{cd} = 17$ N/mm²

Area required $= \dfrac{1206 \times 10^3}{17} = 70941$ mm²

Effective area $\approx 4c^2 +$ Section perimeter $\times c +$ section area

where c is the cantilever outstand of the effective area, as shown below.

$70941 = 4c^2 + 1490c + 9310$

Solving, $c = 37.6$ mm

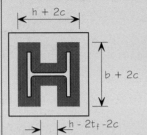

$\dfrac{h - 2t_f}{2} = \dfrac{254.1 - 2 \times 14.2}{2} = 112.9$ mm,
> 37.6 mm

Therefore there is no overlap between the flanges

Thickness of base plate (t_p)

BS EN
1993-1-8
6.2.5(4)

$t_p = c\left(\dfrac{3f_{cd}}{f_y \times \gamma_{MO}}\right)^{0.5}$

$t_p = 37.6 \times \left(\dfrac{3 \times 17}{275 \times 1.0}\right)^{0.5} = 16.2$ mm

$t_p < 40$, therefore nominal design strength $= 275$ N/mm².

Adopt 20mm thick base plate in S275 material

$t_p = 20$ mm

Connection of base plate to column

It is assumed that the axial force is transferred by direct bearing, which is achieved by normal fabrication processes. Only nominal welds are required to connect the baseplate to the column, though in practice full profile 6mm fillet welds are often used.

Unless stated otherwise all references are to BS EN 1993-1-1	<u>Frame stability</u>
	Introduction
	This example examines the building for susceptibility to sway instability (second-order effects). Beam-and-column type plane frames in buildings may be checked for susceptibility to second order effects using first order analysis and the approximate formula:
	$$\alpha_{cr} = \frac{H_{Ed}}{V_{Ed}} \frac{h}{\delta_{H,Ed}}$$
5.2.1(4)B	If $a_{cr} \geq 10$, any second-order effects are small enough to be ignored. The definition of each parameter is given later in this example.
	Figure 12.1 shows the structural layout of the braced bays which are present in each end gable of the building. Unbraced bays occur at 6 m spacing along the 48 m length of the building (i.e. 8 bays in total). The braced bay therefore attracts one half of the total wind loading on the windward face of the building which is assumed to be transferred to the bracing via a wind girder in the roof and diaphragm action in the floor slabs at each floor level.
	The bracing must also carry the equivalent horizontal forces that arise from frame imperfections such as a lack of verticality. The equivalent imperfection forces are based on 1/200 (0.5%) of the total factored permanent and variable load acting on each roof and floor level. These imperfection forces are also distributed to the end bracing via wind girder and floor diaphragm action such that each braced bay receives the equivalent of one half of the total equivalent horizontal force calculated for the whole building.
	<u>Ultimate limit state (ULS)</u>
	The check for susceptibility to second order effects is a ULS check. In this example, the frame will be checked using Equation 6.10b, and only under one load combination with wind as the leading action. In practice, Equation 6.10a would need to be considered, and additional load combinations.
BS EN 1990 Eqn. 6.10b	Design value of actions is $\xi \gamma_G G_k + \gamma_Q \dot{Q}_k + \psi_0 \gamma_Q Q_k$
	Partial factors for actions
BS EN 1990 NA 2.2.3.2 Table NA A1.2(B)	Partial factor for permanent action $\gamma_G = 1.35$ Partial factor for variable action $\gamma_Q = 1.5$ Reduction factor $\xi = 0.925$

| Example 12 Frame stability | Sheet | 2 | of | 4 | Rev |

BS EN 1990 NA 2.2.2 Table NA A1.1	**ψ_0 factors** For imposed floor loads (office areas) $\qquad \psi_0 = 0.7$ For snow loads on roofs (H≤1000m a.s.l) $\qquad \psi_0 = 0.5$	
	<u>Design value of wind load, as the leading action</u> Total wind load on windward face of building = 1.5 × 925 = 1388 kN Total wind load resisted by braced bay = 0.5 × 1388 = 694 kN Distribution : $\quad$ At roof level $\quad$ = 694 / 8 $\quad$ = 86.8 kN $\qquad\qquad\qquad\quad$ At floor levels = 694 / 4 $\quad$ = 173.5 kN	Wind loading on braced bay
	<u>Design value of the vertical loads, in combination with wind as the leading action</u> Roof loading on one braced frame = 14 × 6 [0.925 × 1.35 × 0.9 + 1.5 × 0.5 × 0.6] = 132.2 kN Total roof loading = 8 × 132.2 = 1058 kN	g_k = 0.9 kN/m² q_k = 0.6 kN/m² (see arrangement and actions)
5.3.2(3)	Equivalent horizontal force (acting as a point load) at roof level in end frame = 0.5 × 0.5% × 1058 = 2.7 kN	Equivalent horizontal force at roof level = 2.7 kN
	Floor loading on one braced frame = 14 × 6 [0.925 × 1.35 × 3.7 + 1.5 × 0.7× 3.3] = 679 kN Total floor loading = 8 × 679 = 5433 kN	g_k = 3.7 kN/m² q_k = 3.3 kN/m² (see arrangement and actions)
5.3.2(3)	Equivalent horizontal force (acting as a point load) at each floor level in end frame = 0.5 × 0.5% × 5433 = 13.6 kN	Equivalent horizontal force at each floor level = 13.6 kN
	Note that in accordance with 5.3.2(3) the equivalent imperfection forces may be modified (reduced) by α_h and α_m. It is conservative to ignore these reduction factors. Whereas α_h and α_m reduce the magnitude of the forces transferred to the stability system (in this example, the bracing in the end bays), the effect of α_h and α_m on the value of α_{cr} is modest. In this example, α_h and α_m have been set to 1.0.	

| Example 12 Frame stability | Sheet | 3 | of | 4 | Rev |

Braced Bay Layout

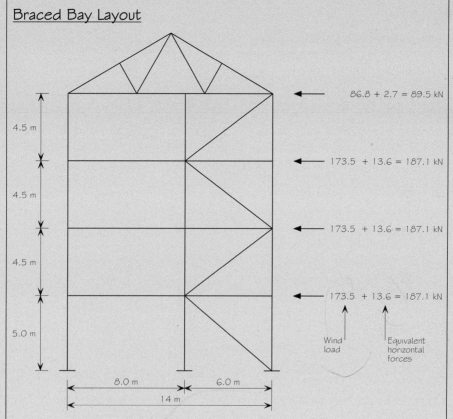

Figure 12.1   Section at braced bay

Assumed Section Sizes

Columns	203 × 203 × 52 UKC
Beams	254 × 146 × 31 UKB
Bracing	219.1 × 10.0 CHS

Sway Analysis

The sway analysis is carried out for horizontal loading only as shown

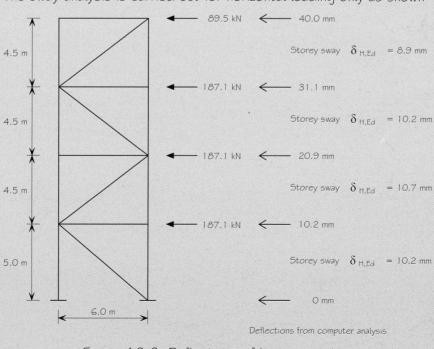

Deflections from computer analysis

Figure 12.2 Deflections of bracing system

| Example 12 Frame stability | Sheet | 4 | of | 4 | Rev |

Assumptions

Column bases both pinned
Columns continuous over full height
Bracing and beams pinned to columns

Frame stability

The measure of frame stability, α_{cr} is verified as follows:

5.2.1(4)B

$$\alpha_{cr} = \frac{H_{Ed}}{V_{Ed}} \frac{h}{\delta_{H,Ed}}$$

where:
H_{Ed} is the (total) design value of the horizontal reaction at bottom of storey
V_{Ed} is the (total) design vertical load at bottom of storey
h is the storey height
$\delta_{H,Ed}$ is the storey sway, for the story under consideration

Fourth Storey:

$H_{Ed,4} = 89.5$ kN

$V_{Ed,4} = 1058 \times 0.5 = 529$ kN

$\alpha_{cr,4} = \frac{89.5}{529} \frac{4500}{8.9} = 85.5 > 10$ — Not sway sensitive

Third Storey:

$H_{Ed,3} = 89.5 + 187.1 = 276.6$ kN

$V_{Ed,3} = 529 + 0.5 \times 5433 = 3246$ kN

$\alpha_{cr,3} = \frac{276.6}{3215} 46 \frac{4500}{10.2} = 37.6 > 10$ — Not sway sensitive

Second Storey :

$H_{Ed,2} = 276.6 + 187.1 = 463.7$ kN

$V_{Ed,2} = 3246 + 0.5 \times 5433 = 5963$ kN

$\alpha_{cr,2} = \frac{463.7}{5963} \frac{4500}{10.7} = 32.7 > 10$ — Not sway sensitive

First Storey :

$H_{Ed,1} = 463.7 + 187.1 = 650.8$ kN

$V_{Ed,1} = 5963 + 0.5 \times 5433 = 8680$ kN

$\alpha_{cr,1} = \frac{650.8}{8680} \frac{5000}{10.2} = 36.8 > 10$ — Not sway sensitive

Therefore, the frame is **not** sway sensitive and second-order effects can be ignored.

7   BIBLIOGRAPHY

7.1   SCI and SCI/BCSA publications

Steel building design: Introduction to the Eurocodes (P361)
The Steel Construction Institute, 2009

Steel building design: Concise Eurocodes (P362)
The Steel Construction Institute, 2009

Steel building design: Design data (P363)
The Steel Construction Institute and The British Constructional Steelwork
Association, 2009

Steel building design: Worked examples – open sections (P364)
The Steel Construction Institute, 2009

Handbook of Structural Steelwork (Eurocode Edition) (P366)
The British Constructional Steelwork Association and The Steel Construction
Institute, 2009

Steel building design: Worked Examples - hollow sections (P374)
The Steel Construction Institute, 2009

Steel building design: Fire resistant design (P375)
The Steel Construction Institute, 2009

Steel building design: Worked examples for students (Without National Annex
values) (P376)
The Steel Construction Institute, 2009

Architectural Teaching Resource Studio Guide – Second Edition (P167)
The Steel Construction Institute, 2003

7.2   Other publications

Steel Designers' Manual 6th Edition
SCI and Blackwell Publishing, 2003

GULVANESSIAN, H., CALGARO, J. A. and HOLICKY, M.
Designers' guide to EN 1990 Eurocode: Basis of structural design
Thomas Telford, 2002

GULVANESSIAN, H., CALGARO, J.A., FORMICHI P. and HARDING, G.
Designers' guide to EN 1991-1-1, 1991-1-3 and 1991-1-5 to 1-7 Eurocode 1:
Actions on structures: General rules and actions on buildings
Thomas Telford (to be published in 2008)

NARAYANAN, R, S. and BEEBY, A.
Designers' guide to EN 1992-1-1 and EN 1992-1-2 Eurocode 2: Design of
concrete structures. General rules and rules for buildings and structural fire
design
Thomas Telford, 2005

GARDNER, L. and NETHERCOT, D.
Designers' guide to EN 1993-1-1 Eurocode 3: Design of steel structures – Part 1.1: General rules and rules for buildings
Thomas Telford, 2005

JOHNSON, R .P. and ANDERSON D.
Designers' guide to EN 1994-1-1 Eurocode 4: Design of composite steel and concrete structures – Part 1.1: General rules and rules for buildings
Thomas Telford, 2004

7.3   Sources of electronic information

Sources of electronic information include:

Access steel web site: www.access-steel.com

Corrosion protection guides – various titles available from Corus web site: www.corusconstruction.com

Eurocodes expert: www.eurocodes.co.uk

NCCI website: www.steel-ncci.co.uk

7.4   Structural Eurocodes

The following Eurocode Parts are applicable for the design of steel-framed buildings, although not all will be required for a specific structure, depending on its use and form of construction.

BS EN 1990 Eurocode – Basis of structural design

BS EN 1991 Eurocode 1: Actions on structures

BS EN 1991-1-1	Part 1-1: General actions. Densities, self-weight, imposed loads for buildings
BS EN 1991-1-2	Part 1-2: General actions. Actions on structures exposed to fire
BS EN 1991-1-3	Part 1-3: General actions. Snow loads
BS EN 1991-1-4	Part 1-4: General actions. Wind actions
BS EN 1991-1-5	Part 1-5: General actions. Thermal actions
BS EN 1991-1-6	Part 1-6: General actions. Actions during execution
BS EN 1991-1-7	Part 1-7: General actions. Accidental actions

BS EN 1992 Eurocode 2: Design of concrete structures

BS EN 1992-1-1	Part 1-1: General rules and rule for buildings
BS EN 1992-1-2	Part 1-2: General rules. Structural fire design

BS EN 1993 Eurocode 3: Design of steel structures

BS EN 1993-1-1	Part 1-1: General rules and rules for buildings
BS EN 1993-1-2	Part 1-2: General rules. Structural fire design
BS EN 1993-1-3	Part 1-3: General rules. Supplementary rules for cold-formed members and sheeting
BS EN 1993-1-5	Part 1-5: Plated structural elements

BS EN 1993-1-8	Part 1-8: Design of joints
BS EN 1993-1-9	Part 1-9: Fatigue strength
BS EN 1993-1-10	Part 1-10: Material toughness and through-thickness properties
BS EN 1993-1-12	Part 1-12: Additional rules for the extension of EN 1993 up to steel grades S700

BS EN 1994 Eurocode 4: Design of composite steel and concrete structures

| BS EN 1994-1-1 | Part 1-1: General rules and rules for buildings |
| BS EN 1994-1-2 | Part 1-2: General rules. Structural fire design |

National Annexes

UK National Annexes are published by BSI.

Published Documents

PD 6695-1-10:2008
Recommendations for the design of structures to BS EN 1993-1-10
PDs are available from BSI

7.5 Product Standards

BS EN 10025-2:2004 Hot rolled products of structural steels. Part 2: Technical delivery conditions for non-alloy structural steels

BS EN 10164:1993 Steel products with improved deformation properties perpendicular to the surface of the product. Technical delivery conditions

BS EN 10210-1:2006 Hot finished structural hollow sections of non-alloy and fine grain structural steels Part 1: Technical delivery requirements

BS EN 10219-1:2006 Cold formed hollow sections of non-alloy and fine grain steels. Part 1: Technical delivery conditions

Typeset and page make-up by The Steel Construction Institute, Ascot, Berks. SL5 7QN
Printed by Information Press, Eynsham, Oxford OX29 4JB
1,200   06/09   PUB8010